THIN ON THE GROUND

SECOND EDITION

THIN ON THE GROUND

Soil Science in the Tropics

Second Edition

Anthony Young

Land Resources Books

Copyright © 2007, 2017 Anthony Young.

Anthony Young has asserted his right under the Copyright, Designs and Patents Act 1988 to be identified as the author of this work.

Second Edition published in 2017
by Land Resources Books, Norwich, Great Britain.

First published in 2007
by The Memoir Club, County Durham.

A CIP catalogue record for this book
is available from the British Library.

ISBN-10: 0995656606
ISBN-13: 978-0995656604

In gratitude for all her support to my career and for the joys of a marriage of 60 years this book is dedicated to Doreen.

Other books by the author

Slopes
Tropical soils and soil survey
A framework for land evaluation (with Robert Brinkman & FAO)
Soil survey and land evaluation (with David Dent)
Agroforestry for soil conservation
Guidelines for land use planning (with FAO)
Agroforestry for soil management
Land resources: now and for the future
Semper Juvenis: Always Young (Autobiography)

Books for schools

A geography of Malawi (with Doreen Young)
World vegetation (with Dennis Riley)
Slope development (with Doreen Young)

CONTENTS

LIST OF ILLUSTRATIONS

Unless otherwise acknowledged, photographs are by the author.

Pioneers
Original staff of the Imperial College of Tropical Agriculture, Trinidad, 1922; Frederick Hardy standing left. (*Reproduced from The Imperial College of Tropical Agriculture: Silver Jubilee, ICTA, Trinidad, 1951*) (p.18)

An early account of the environmental damage caused by vegetation clearance, in the somewhat austere format of Government publications of the day. (p.23)

Arthur Hornby's Soil Map of Central Nyasaland (1935), covering what are now parts of Southern and Central Regions, Malawi. The original is in brilliant colours at scale 1:350 000. (p.25)

A. W. R. Joachim. (*R.O.B. Wijesekera*) (p.28)

Geoffrey Milne. (*Reproduced from Bibliographical Studies of Geographers, ed. T. W. Freeman and P. Pinchemel, Volume 2, 1978*) (p.30)

Colin Trapnell and Neil Clothier on safari in Zambia. (*Royal Botanic Gardens, Kew, communicated by Paul Smith*) (p.37)

East Africa
Cliff Ollier explains the purposes of soil survey to the Karamajong people, Uganda. (*Cliff Ollier*) (p.68)

An early air photograph: the Gezira Research Farm, Sudan, 1929. (*Reproduced from Agriculture in the Sudan, J.D. Tothill, 1948, by permission of Oxford University Press*) (p.84)

Distinguished visitor: Dr H. Kamuzu Banda, shortly to become Life President of Malawi, being shown air photographs of Livingstonia Mission, where he received his early education; Anthony Young centre. (p.163)

South Asia: Canal irrigation on the upper Indus, Pakistan

Survey with armed guard, near the northern frontier of Pakistan. (*Ian Twyford*) (p.181)

Large-scale irrigation: constructing concrete liner tubes for the Tarbela Dam on the Indus, the largest earth-filled dam on the world, completed in 1976. (p.182)

Canal headworks: from the River Indus to the major canal. (p.183)

Minor canal. (p.183)

From a minor canal to farmer's land: measuring water flow through a nakkar. (p.184)

The end point: water reaches a cotton crop via a field channel. (p.184)

South-east Asia: survey for the Jengka Triangle Project land settlement scheme, Malaysia

One of three survey teams ready to set off into the forest; Anthony Young centre; the *mandor* (foreman), Ramli bin Puteh, holds the kettle for making tea. (p.195)

Crossing the River Pahang (with limited freeboard). (p.196)

Soil survey in rain forest: neither air photographs nor view of the landscape are available. (p.197)

Logging company clearing the forest. (p.198)

The project under way, based on land use planning with conservation agriculture. Foreground, oil palm with cover crop on gentle slopes with rich soils. Middle distance, 100% protection of sloping land with cover crop, prior to planting rubber. Background, natural *Dipterocarp* forest retained on summits to reduce runoff. (p.199)

Soil erosion and land degradation

Gullying in a valley floor (*dambo*), Malawi; the gullies lower the water table, hence dry-season pasture is no longer available. (p.229)

Salinization of irrigated land, Pakistan. (p.229)

Severe, irreversible erosion in Haiti. This hillslope was covered by a conservation programme of contour hedgerows promoted by an aid agency; the foreground shows consequences when a storm breaches the conservation works. (p.232)

Talking with farmers, World Bank Sokoto State Agricultural Development Programme, 1982. (p.236)

Reunion of soil surveyors, Royal Geographical Society, 2010; from left Maurice Purnell, Helen Sandison (née Brash), Hugh Brammer, Anthony Young. (p.257)

Using survey results internationally

FAO consultation for a project on Population-Supporting Capacity; at table from left, A. P. A. Vink (Netherlands), Jean King (UK, Unesco representative), Graham Higgins and Anthony Young (UK); behind: Maurice Purnell and Amir Kassam (UK), Rudi Dudal (Belgium). (*Graham Higgins*) (p.266)

Soil surveys by year of publication

Holdings of the World Soil Survey Archive and Catalogue (WOSSAC). (*Adapted from data in www.wossac.com by permission of Stephen Hallett*) (p.277)

Population

Farmers in Malawi forced, through population increase, to support their family on less than half a hectare. Because there is no spare land, cultivation has been extended onto the hill area in the background. (Cover image)

FOREWORD

Professor Stephen Nortcliff
Secretary General, International Union of Soil Sciences 2003-2010

From my position it has been a rewarding experience to see Soil Science rising up the agenda of institutions concerned with the environment and sustainable development. Whilst to soil scientists this is only a statement of the obvious, in recent years it has taken a great deal of effort to achieve such recognition. In Europe this progress has been marked by moves towards legal frameworks for the protection of soils, and at national level many countries have taken steps to promote sustainable management of the land resources on which our welfare depends.

In the developed world, management of the environment has been based on knowledge accumulated over many years, a key part of which were the maps produced by national soil survey organisations. In developing countries, whilst farmers possessed much wisdom about their lands under traditional methods of management, there was no systematic background of scientific knowledge on which to base the developments planned after the Second World War. This was a critical era in the development of soil and land resource surveys. Their rationale was a recognition that knowledge of the natural environment was essential if sound decisions were to be made on future developments.

In this text, Anthony Young seeks to recognise the pioneering efforts of those who undertook the surveys. At the time they would probably not have called themselves environmental scientists, but in retrospect they were the predecessors of what we now recognise as such, seeking to understand the nature and distribution patterns of the landscapes which they observed in the field. The information collected during these surveys was often tremendously detailed. Unfortunately its true worth was not fully recognised at the time, and much was produced in documents of limited local circulation. Whilst there have been recent projects to record the information gathered during these surveys, these have been on a limited scale and much information has been lost, or will be if action is not taken soon.

Within the United Kingdom, many young scientists gained their first practical experience as part of these land resource teams. When I first began to work on the soils of Norfolk I was fortunate to be supervised by Young who had started his career in the Colonial Service in Central Africa. I also had close contacts with colleagues in what was then the Soil Survey of England and Wales. They enlightened me about soil-landscape relationships and the often intricate patterns of soil distribution. What I did not realise at the time was that many of these individuals had gained their first practical insight into soils through involvement in tropical and subtropical regions.

This book is the story of a great era of land resource surveys, emphasising the importance of the experiences gained from observing the natural environment in the field. It also recognises the contribution of these scientists to the history of the Colonial Service. In gathering information from some 80 former colleagues, the author asked them to include reminiscences of unusual experiences in the field, which provide lighter moments in the text.

In a retrospective look at the work done during this period, Young asks whether the objectives of these surveys were clearly elaborated at the time, and if so, were they achieved. What mistakes were made, particularly in communicating the results to governments and development agencies? Given the enormous amount of material collected during these surveys, a critical appraisal made of subsequent development plans shows without doubt that this wealth of information was not always used effectively and was sometimes left to gather dust. Why was this the case? What can development institutions, and the governments of independent countries, learn from the way natural resource data were used, or under-used, in the past? With the contemporary focus of development projects on poverty reduction and sustainable development, it is essential that the lessons learnt from the past be incorporated in these new initiatives. Anyone currently undertaking a development project in these regions should search the archives to seek out these earlier surveys.

Anthony Young is to be congratulated in bringing together information about the surveys, and recording the careers and achievements of the surveyors. This early work, carried out initially under the auspices of the Colonial Office and subsequently by the UK-based Land Resources Division, provides much of the basic knowledge we have about the natural environments of many of the regions today. This text will help to bring about a wider awareness of the wealth of information collected during this era.

Stephen Nortcliff
University of Reading

PREFACE TO THE SECOND EDITION

Over a period of 30 years, 1950-1980, the greater part of agricultural land in the tropics, actual and potential, was covered by surveys of natural resources: soil, climate and vegetation. It can be called the golden age of soil survey. How did this remarkable achievement come about, and what lessons does it hold for us today?

The objectives remain similar: to give an account of how the major advances in knowledge of tropical land resources and their management were accomplished, and to examine the legacy of these achievements for soil science today. In the First Edition the emphasis was on the story itself. In this Revised Edition the focus is on how the experience gained in the past has lessons for the study of soils in the field today.

This book describes the achievements for countries of the British Commonwealth. Similar accounts could be written of surveys in former French territories of Africa, and advances in tropical soil science made by Dutch scientists working in south-east Asia.

The regional accounts are largely unchanged. Four countries, Ghana, Tanzania, Sudan and Swaziland, have been updated using newly available sources. Two more biographies have been added, of Anthony Smyth and Maurice Purnell, both of whom began their careers in soil survey and went on to take leading posts with international institutions. A colleague when looking at the proof copy remarked that the book was "dense with information", so an Index of Personal Names has been added and the General Index much expanded.

The account of soil conservation has been rewritten and expanded. What began as concern for soil erosion, and the failure of attempts to impose protection by coercion, evolved into a broader appreciation that conservation comprised retaining the soil's physical, organic and chemical properties, and thereby its sustained productivity. Erosion control became, successively, soil conservation in this wider sense, then land husbandry and conservation agriculture.

Also expanded and updated is the account of attempts to assess relative degrees of soil and land degradation, initially on a controlled subjective basis and subsequently based on quantitative but indirect methods. Since the 1990s, monitoring of changes in soil conditions, sometimes called soil health, has become a leading concern.

The final chapter is new. It is clear that, by comparison with the past, very few soil surveys are being conducted today. A high proportion of soils research is based on laboratory studies. This leaves the author with a feeling of unease. So what should be the role today of the study of soils in the field? Two aspects are identified: monitoring of soil health based not only on sampling and analysis but on interacting with the experience of farmers; and the value of soil scientists as specialist support to agricultural advisory services.

The history of soil survey given in the First Edition could not be written today. I gained much from the personal recollections, dating back in some cases more than 60 years. Similarly, in describing of how conservation agriculture evolved I have been fortunate to meet many in those who took leading parts. I hope that by setting out the achievements of the past, as well as the mistakes made, this account can contribute to a the role of field study of soils today.

Anthony Young
Norwich 2017

PREFACE TO THE FIRST EDITION

An era of development in territories of the British Commonwealth took place following the Second World War. It was set off by a change in attitudes. Initially, there was a heightened concern for the expansion and security of world food supplies. This gave way to thinking in terms of development as it is now understood, improvement in the welfare of peoples in the Third World.

It was clear that initially, development had to come from the rural sector, mainly agriculture. This in turn rested on land resources: climate, water, soils, pastures and forests. Apart from worthy but isolated pioneering efforts between the wars, very little was known about the nature and distribution of these resources. There were no reliable maps of soils and vegetation, nor knowledge of their potential for improved land use. The farmers, of course, knew about their own land, and local agricultural officers built up knowledge of their districts; but there was no systematic knowledge based on a common scientific language, that would allow exchange of experience in land management, and provide a basis for development planning.

Realizing this, the British Government recruited Soil Surveyors and Ecologists, mainly recent graduates, young and enthusiastic. They set off into the bush to make maps of the land resources and assess their development potential. When former Colonial territories achieved independence, the work was taken over by the Land Resources Division, a British aid organization which seconded staff to where they were needed. From 1950 to the early 1980s large areas, in some cases whole countries, were covered by reconnaissance surveys. What had not been known before was now on record: the land resources of the British Commonwealth.

This is the story of that era, and of the pioneering efforts that preceded it. One objective of this book is to form part of the history of science, not based on experiment but on observation of the natural environment in the field. A second is to make a contribution to the history of the Colonial Service, most previous accounts of which have been written by administrative officers. When preparing this account, I contacted many former colleagues for information and recollections, and was impressed by their dedication. What started as a description of the surveys became, more and more, the story of the surveyors. They achieved much in a short space of time. Those who subsequently left field survey went on to careers with international and British development institutions and in universities. Their early experience of the complex interactions within the tropical environment seemed to have conferred both an enthusiasm and a springboard of knowledge which they never lost.

There is a third objective: the lessons to be learnt for the future development of Britain's former overseas territories, now largely independent. What were the intentions of the surveyors, and were they achieved? The scientists who described the land resources did not always make a good job of putting across their results, particularly their significance for land management. Partly for this reason, and partly because the pendulum of opinion has currently swung towards a sociological approach, many development agencies have lost sight of fundamental facts about the potential and limitations of the natural environment. Experience gained in the past will help present-day development agencies and national government institutions to work more effectively towards reducing hunger and poverty, conserving the environment, and improving the welfare of their people.

Anthony Young
Norwich 2007

ACKNOWLEDGEMENTS

A large number of soil surveyors, ecologists and other land resource scientists responded willingly, often enthusiastically, to my request for information. They sent details of their education and careers, main contributions to field survey, and the use which was made of their work. Many also sent anecdotes of unusual or bizarre experiences in the field. Some sent up to a dozen instalments or replies to enquiries, but it would be invidious to indicate these, as all were as helpful as they could be. These are the colleagues to whom I extend warm thanks; without their help this book could not have been written.

John Adams
Nazeer Ahmad
Peter Ahn
Brian Anderson
Gordon Anderson
R. D. Asiamah
Ian Baillie
Colin Baker
John Baulkwill
John Bennett
Len Berry
David Billing
Pieter Bleeker
Hugh Brammer
Helen Brash
Robert Brinkman
Peter Brown
Martin Brunt
Jeff Burley
Marion Cheatle
Allan Chilimba
John Coulter
Dilip Pal

Sandie Crosbie
Barry Dalal-Clayton
Martin Doki
Malcolm Douglas
William Effland
David Eldridge
Henry Elwell
Herbert Farbrother
John Goldsack
Robert Green
Dennis Greenland
Richard Grove
John Hansell
Alfred Hartemink
John Harrop
Douglas Helms
Graham Higgins
Priscilla Higgins
Ian Hill
John Howard
Geoffrey Humphreys
Charles Johnson
Robert Ridgway

Alun Jones
Amir Kassam
Anthony Kirk-Greene
Karl Kučera
Rick Landon
Ian Langdale-Brown
J. G. Lyimo
John McAlpine
Heather McAvoy-Marshall
John Makin
Ranjith Mapa
Anthony Mitchell
Colin Mitchell
David Moffat
Roy Montgomery
John Morris
George Murdoch
Ng Siew Kee
Kingston Nyamaphene
Peter Nye
Cliff Ollier
Chris Panabokke
Roel van der Weg

David Parry	Vernon Robertson	William Verboom
Garry Paterson	Francis Shaxson	Peter Vine
Ron Paton	Paul Smith	Keith Virgo
Jim Perry	Anthony Smyth	David Wall
Christine Kowal Post	Michael Stocking	Richard Webster
Maurice Purnell	Bernard Swan	Marina Wijesekera
David Radcliffe	Colin Trapnell	Alan Yates
Estela Radwanska	Ian Twyford	

These replies have been deposited with the World Soil Survey Archive and Catalogue (WOSSAC).

I should like to extend thanks to the most helpful library staff at the British Library, London, Cambridge University Library, Rhodes House Library, Oxford, and the World Soil Survey Archive and Catalogue, Cranfield University. Also to staff at the Natural Resources Institute for information on the Land Resources Division, the Overseas Pensions Department of the Department of International Development, and Anthony Kirk-Greene for advice on Colonial Office staff lists. It was at the suggestion of Dan Yaalon of Israel, doyen of the history of soil science, that I embarked on this project.

In preparing the Second Edition I am grateful to the following for supplying new material and for comment and suggestions: Wayne Borden, David Dent, Stephen Hallett, Heather McAvoy-Marshal, Robert Ridgway and Francis Shaxson.

Finally, I should like to acknowledge the considerable editorial help received from Howard Smith, in casting an eagle eye over my drafts, styling the text, processing the illustrations and overseeing the publication process from start to finish. My thanks also to Stephen Low who spent many hours scanning slides and prints from my archive.

COUNTRY NAMES AND INSTITUTIONS

Many of the countries covered changed their names on reaching independ-ence. In the First Edition the pre-independence names were employed to help convey the flavour of the period. It is now over 50 years since these became obsolete, and in this Edition the modern names are used, with the earlier name added on first appearance, e.g. 'Malawi (Nyasaland)'. Excep-tions are in cases where the pre-independence names are used in quotations or names of institutions.

FAO refers to the Food and Agriculture Organization of the United Nations, with Headquarters in Rome. Two institutions changed their names several times. 'Land Resources Division' is used throughout to refer to the British aid organization which in 1978 became the Land Resources Devel-opment Centre, and subsequently a Department within the Natural Resources Institute. 'Huntings' is used as an abbreviation for Hunting Tech-nical Services Ltd. and other companies within the Hunting group. See also Index of Acronyms (p.318).

1

SETTING THE TASK

> The distribution of the great soil belts over the continent of Africa
> seems to have received hitherto no attention whatever.

THIS STATEMENT COMES FROM *The Vegetation and Soils of Africa*, published in 1923 by the American Geographical Society. The section on soils was written by the eminent scientist Curtis Fletcher Marbut, Director of the US Bureau of Soils, and is significant for what it did not, and could not, describe. Marbut found that, apart from a little work in South Africa, there was almost no information available. Attempts to draw a world soil map prior to 1950 rested upon inference from the broad climatic belts. Whilst farmers possessed a wealth of knowledge about their own soils in their immediate area, there was no systematic, global knowledge of soil types and their distribution.

This can be contrasted with the situation in 1964, when the Inter-African Pedological Service, under the direction of J.L. d'Hoore of Belgium, completed the first *Soil Map of Africa*. The map, in seven sheets at a scale of 1:5 000 000, shows soil units in some detail. Its compilation required massive approximations, interpolations, and for some areas intelligent guess-work, but at this broad reconnaissance scale the distributions mapped are clearly recognisable by anyone with a knowledge of the countries concerned. This is because d'Hoore was able to call upon the help of over 100 soil specialists working in forty-six countries. The same situation was found in Asia, Central America, and countries of the Pacific. By 1961, when the FAO began to compile a *Soil Map of the World*, they were able to base it upon national soil maps compiled by survey organisations in nearly every country. Thus in some forty years, knowledge about the soils of the developing world was transformed.

The greater part of this book is an account of that transformation: how it was brought about, the institutions through which it was effected, and the soil surveyors, ecologists and other field scientists who achieved it. The change took place in two phases: scattered pioneering work from 1920 to 1939, and a golden age of reconnaissance surveys, conducted as part of the international drive for development after the Second World War, from 1950 to about 1980. Since then, national efforts have become focused upon more detailed surveys for immediate development planning purposes, whilst international activities have moved away from survey as such to land evaluation and land use planning.

In parallel with survey is the story of soil conservation, how the early method based on earth structures evolved into the modern approach of conservation agriculture. The time scale here is more extended, beginning as land husbandry from the 1960s onwards, and still gaining support into the 21ˢᵗ century.

The scope of this book

The scope of this account can be set in three respects: area, time, and resources. The area coverage is the tropical and subtropical, less developed, parts of the British Commonwealth, comprising some 40 countries which were formerly colonies, together with India. The Dominions of the developed world, Australia, New Zealand, South Africa and Canada, each of which established outstanding national survey organisations, are excluded. Parallel accounts could be written of the considerable amount of survey work conducted in the former French territories of Africa, Belgian work in the Congo, and Dutch work in their overseas possessions, primarily the East Indies.

After a brief mention of accounts of land resources by early explorers, the time range begins with the surveys in the 1920s and 1930s, and ends at about 1980, when reconnaissance surveys were largely completed, most colonies had achieved independence, and the staffing of survey institutions had changed from being predominantly British to local. The terminal date is treated flexibly, indicating where soils research has been taken up by locally-staffed institutions.

Most of the studies described were called soil surveys. Other terms employed in more or less the same sense are ecological or agro-ecological surveys. Soil survey is a somewhat misleading term. Every soil survey is a land resource survey, where land resources are the properties of the natural environment which provide the basis for agriculture and other kinds of rural land use: geology, landforms, climate, hydrology, soils, vegetation, and occasionally fauna. Indeed, in many reconnaissance surveys, the maps do not show soils, but land systems or other units of the physical environment as a whole.

Why carry out land resource surveys?

The most basic reason to complete land resource surveys is to avoid disasters. The classic case is the East African Groundnuts Scheme of 1947, an ambitious UK-financed development project. Three areas in Tanzania were selected for mechanized vegetation clearance and planting of ground-nuts. At one, the rainfall was too low and the soils excessively compact and abrasive on machinery; at another, much land was subject to waterlogging. The story of this scheme is told in Chapter 4. As a result of this costly failure it became widely recognized that developments which involve changes in land use should be preceded by resource surveys, to select the most suitable areas and warn of hazards. This soon became a standard requirement for internationally funded projects, but it applies equally to government or private investments.

Farmers, of course, know about their land so far as the traditional crops and management practices are concerned. Experienced agriculturalists and foresters, working in a country over many years, build up a store of know-ledge, although this is not always transferred to their successors. Local knowledge is not enough, however, when new crops are introduced. If tea cultivation, for example, is extended from estates to smallholder production, surveys are needed to ensure that the crop is not promoted on unsuitable land.

In a general sense the objective of all surveys is to serve as an important basis for decisions on land use and management. In 1954 Ernest Chenery, when putting forward his proposal for a soil, vegetation and agricultural survey of Uganda, expressed it as, 'A background to assessing the present and future agricultural potential of the Protectorate, and its viability for capital investments … [and specifically] the allocation of foreign aid.' Hugh Brammer, after his many years of work on soil and land evaluation surveys in Bangladesh, summarized their purpose as, 'To identify different soils, to describe and map them, and to evaluate their properties in relation to land use and management.'

These are broad, overall objectives. To become more specific, surveys can be grouped into three types: reconnaissance, project, and detailed. Dividing them on the basis of purpose is better than using scale as a basis, since reconnaissance surveys of small countries may be on scales more typically found for project surveys. What follows are the potential, and desirable, applications. Whether they have in fact been used in this way is considered in Chapter 15.

Reconnaissance surveys extend over the whole of a country or large region. Typically they are at scales between 1:250 000 and 1:500 000. These are general-purpose surveys, not aimed at any specific type of land development, and usually cover the whole spectrum of land resources: climate, hydrology, landforms, soils and vegetation. In a very general sense, their aim is to show what is there. Specifically, the aims of reconnaissance surveys are:

- To show the extent of land suited to different types of use (agriculture, forestry, wildlife reserves) and to particular crops. This makes the survey a starting point for project identification, including choice of areas.

- To provide a national framework of soil types. This is for future use in project and detailed surveys, and also as a basis for the two following applications.

- To select sites for applied agricultural research: experiments, crop variety trials, fertilizer trials. There are two sides to this: selecting sites for the trials, and identifying the areas over which their results are applicable, sometimes called recommendation domains.

- As a basis for agricultural extension. It is certainly not possible to say, 'The map shows the Tajong series here, so we should apply management practices suited to that type of soil', because the high spatial variability of soil properties cannot be shown on maps at reconnaissance scale. However, through field visits by a soil scientist and agricultural officer together, or by training workshops, extension staff can be shown firstly, how to recognize the local soil types, and secondly, what are the management needs of each. They do not need detailed, field by field, surveys; once the extension worker and the farmer establish that they are talking about the same kind of soil, the latter knows very well where it is to be found.

- To evaluate the service roles of land resources, for example water catchments, conservation of wildlife and genetic resources, promotion of tourism.

- As a background to studies of other aspects of the rural sector. This includes research into soils as such, into their relations with vegetation, water, and wildlife, and in an ideal world, economic and social work.

- Educational uses. Agricultural or forestry officers going to work in an unfamiliar region need to be instructed in its climate, soil types and land management needs. In Colonial times this applied to expatriates. Today it is equally more necessary for local staff, many of whom will start with just as limited a knowledge of their own country.

- As an input to land resource databases. This application arose in more recent years. A number of developing countries established digitised databases of land resources, including climate, soils, vegetation and, sometimes, land use and land status (e.g. forest reserves). By adding crop requirements or other needs for land use, digitised land evaluation databases were produced. The soil survey maps are converted to digital form as a primary input to these databases.

Project surveys cover land provisionally intended for a development scheme. They are usually at intermediate, semi-detailed scales such as 1:100 000 or 1:50 000. This type of work is usually carried out before the project has been firmly approved and planned, often called the pre-feasibility stage. The objectives of project surveys are relatively straightforward: hazard avoidance, project design, and project viability.

Hazard avoidance, mentioned above with respect to the East African Groundnuts Scheme, is needed to avoid costly mistakes in development.

This requires identification of the types and location of hazards, such as drought (soils with low water-holding capacity), land liable to erosion, or salinity. Having overcome this hurdle, the next objective is project design, or land use planning. The design of a project calls for setting out land suited to different kinds of use — agriculture, forestry, grazing, wildlife reserves, etc. — and information needed to select land suited to specific crops. This type of application is required when taking new land into cultivation, and particularly for irrigation projects. Developments of this kind were common in the post-war period when 'spare land', cultivable but uncultivated, was plentiful. Under present-day conditions of land scarcity, new land settlement schemes are not often found.

The third objective is to provide information needed to determine project viability, and sometimes choice between two alternative projects. Although environmental considerations also play a part, for better or worse such decisions are made primarily on economic grounds, using cost-benefit analysis, and this requires quantitative assessments of inputs and production. Costs of inputs are estimated on the basis of the management requirements of each soil type: for agriculture, conservation works, drainage, fertilizers, and other inputs. To estimate the quantity of outputs requires information on expected crop yields, livestock production, or forest products. It has to be added that development projects frequently have not made full use of the potential of land resource information.

Detailed surveys are for specific local developments: an oil palm plantation, a sugar estate, or a new agricultural experiment station. They are usually on semi-detailed or detailed scales, typically 1:25 000 or 1:10 000. Surveys at such scales are valuable for commercial crop production, whether on estates or by smallholders, but are not usually justified for less intensive agriculture. They have much the same objectives as project surveys: site selection, hazard avoidance, and internal layout.

It will be apparent that all these applications call for information over and above soil maps. To make surveys useful, they must be extended by land evaluation in its widest sense. Whilst it is possible for this to be a subsequent stage, it is very much better to make it an integrated part of the soil survey. In this way the soil properties that matter, perhaps texture, acidity or salinity, can receive special attention during the mapping. To appreciate what economists are going to ask for, soil surveyors need to be trained in the essentials of economic appraisal. They may even need to argue on such lines as, 'Yes, I can see that your cost-benefit analysis is favourable, but do you realize that in ten to twenty years' time, much of this land will be eroded/degraded/salinized, and what happens to the people then?'

There are other important applications of soil surveys which, regrettably, were rarely to be found during the post-war period. The first of these is in ongoing land management, once a project is under way. There should be a soils expert, ideally one of the original surveyors, as one of the specialists available for consultation when problems arise, who can give advice on modifications to soil management. The second ongoing application is soil monitoring, the observation of changes in soil fertility, and in particular of erosion and other forms of soil degradation. It was not until the 1990s that this was advocated as a basic task for national survey organisations, and only now is it being put into practice. More will be said on these potential, but largely unrealised, applications of surveys in Chapter 16.

Soil science in the United Kingdom ...

Britain's long and distinguished record in soil science was based on practical land management by farmers. The mediaeval three-field system and the Norfolk four-course rotation (wheat, barley, a legume, fallow), introduced in the eighteenth to early nineteenth centuries, were crop rotations designed to maintain fertility — today this would be called sustainable land use. The first to describe a soil profile was Charles Darwin in a paper of 1837, and his book on earthworms (1881) contains many examples.

Experimental agriculture in Britain dates from the foundation of Rothamsted Experimental Station in 1843. In the same year the Broadbalk field experiment was established which is still being maintained today, the longest continuous field experiment in the world.

In 1912 the Goldsmiths' Company Soil Chemist at Rothamsted, E.J. Russell, produced a book, *Soil Conditions and Plant Growth*, a volume in a series, *Monographs on Biochemistry*. From this modest beginning, it was to continue into twelve editions, the first seven by Sir John Russell as he became, the eighth to tenth revised by his son, Walter Russell, and the eleventh and twelfth by a consortium of staff, mainly from Reading University, edited by Peter

Gregory and Stephen Nortcliff. The title is significant. Soils were studied not primarily as natural bodies, as was the case in the early work of Russian scientists, but as the basis for plant growth in agriculture.

Sir John Russell remained active until late in life. On his ninetieth birthday in 1962 he was an honoured guest at the annual British Society of Soil Science conference. Speaking rapidly, unlike most elderly people, he said how delighted he was to find so many soil scientists. When he was a young man, such a meeting would have numbered no more than ten. He still hoped to finish his *History of Agricultural Science in Britain*, which he did, completing it a few weeks before his death in 1965.

The precedents in mapping soils are no less distinguished. The origin of soil survey is generally attributed to the foundation of the United States Bureau of Soils in 1901. However, many years earlier, from 1793 to 1813, Arthur Young, Secretary to the Board of Agriculture, London, produced a series of reports on English and Scottish counties, with the titles, *General View of the Agriculture of the County of* ... Ten were produced 1793-94, continuing with additional counties and revised editions to 1813. They contain maps of soils and types of farming, showing distributions easily recognisable in the light of present-day knowledge. The value of these surveys was evidently recognized at the time, King George III writing, 'Mr Young, I am more obliged to you than to any other man in my Dominions.'

A century later E.J. Russell with the Director of Rothamsted, A.D. Hall, wrote *A Report on the Agriculture and Soils of Kent, Surrey, and Sussex* (1911), a full-length book. Besides detailed accounts of the soils associated with each geological formation, they produced a series of dot maps ('each dot represents x acres') showing distributions of arable, grass, woodland, cattle, sheep, and eight crops; these are printed on transparent paper so that they can be overlain onto a geological map at the same scale to show the correspondences, a remarkable anticipation of modern geographical information systems.

In the 1930s G.W. Robinson, working from the University of Wales in Bangor, more or less single-handedly founded the Soil Survey of England and Wales. The Soil Survey of Scotland operated from 1938 to 1986. An early figure of importance was G.R. Clarke of Oxford University, whose book, *The Study of the Soil in the Field* first appeared in 1936 and ran to five editions. Clarke's wisdom was also transmitted to students in tutorials, some held in the bar of a local hostelry.

...and in the tropics

The foundations of tropical soil science in the British Commonwealth were laid by Frederick Hardy, appointed Soil Chemist at the Imperial College of Tropical Agriculture, Trinidad, in 1920. His research into soil-crop relationships, passed on to successive cohorts of students taking the Diploma of Tropical Agriculture, was of inestimable value in spreading awareness of the role of soils. More about the work of Hardy and other outstanding pioneers is described in Chapter 2.

The starting point for widespread interest in soils of the colonies was the Imperial Agricultural Research Conference held in 1927. Decisions taken at this conference led to the setting up of eight Imperial Bureaux. Of these, the Imperial Bureau of Soil Science was opened in May 1929 at Rothamsted Agricultural Experiment Station under the direction of Walter Russell. This was renamed the Commonwealth Bureau of Soil Science, and subsequently became part of CAB International. In these days of mass circulation of abstracts, it is of interest to see the germ of this practice, monthly cyclo-styled *Bulletins* of the Bureau, sent around the colonies, each reporting some twenty items of research.

Such were the foundations of soil survey in the tropical and subtropical countries of the British Commonwealth. Lest it be thought that contributions from other countries, of equal magnitude and distinction, are being ignored, a few pointers may be given. Dutch scientists in Indonesia (formerly the East Indies) produced a large volume of early work. An obituary of E.C. Julius Mohr, who died in 1970 at the age of ninety-six, considered that, 'In 1906 he became the indisputable leader of tropical soil science'. His four-volume work on soils of the tropics, produced 1934-1938, became more widely known when it was translated by Robert L. Pendleton in 1943. Another early book on tropical soils was by the German, Paul Vageler, translated into English by Herbert Greene in 1933.

The earliest soil map from the tropics was produced by the Dutch, a survey of an area in Sumatra dating from 1901. This was closely followed by the first overseas survey carried out by the US Bureau of Soils in 1902, a sixteen kilometre-wide strip from north to south across Puerto Rico. The earliest from British territories were surveys of two West Indian islands, Dominica and Montserrat, by Frederick Hardy in 1922.

The French overseas research institution, the Office de la Recherche Scientifique et Technique d'Outre-Mer (ORSTOM)[1] was founded, remarkably, during the Second World War, with a focus of work on its North and Central African territories. Communication between French and English-speaking scientists was helped by the Interafrican Pedological Service, founded by the Belgian, J.L. d'Hoore, and located in the then Belgian Congo until political disturbances forced its closure. Its achievements included the first *Soil Map of Africa* in 1954, and a journal *Sols Africains/African Soils*. These strands of research interacted with work in British Commonwealth territories.[2]

Phases of soil survey

Three phases of soil survey in the tropics can be distinguished:

- The pioneering phase, 1920-1939. A small number of far-sighted individuals demonstrated the potential of ecological and soil surveys, and how they could serve land management and development.

- The reconnaissance phase, 1950-1980. This was the golden age of land resource surveys. Work was done at a wide range of scales: reconnaissance surveys of whole countries or large regions; studies at intermediate scales for development projects, including new land settlement and irrigation schemes; and detailed surveys for local purposes, such as experimental farms or plantations. Work during this period was conducted largely by British expatriate staff, initially in the Colonial Service and later by the Land Resources Division.

- The consolidation phase, from 1980 onwards. Land development planning continued, but as most areas suited for use were already taken up, the advisory purpose, helping farmers to make the best

---

[1]   Now part of the Institut de Recherche pour la développement.

[2]   Summaries can be found in the abstracts of a symposium, *History of Soil Science in Developing Countries*, held at the 18th World Congress of Soil Science, Philadelphia, July 2006.

use of their land, came to the fore. The consolidation phase approximately coincides with achievement of independence by former colonies, to be followed by a transition from expatriate to local staff.

The pioneering phase is covered in Chapter 2, taking as a basis the small number of individuals who had the foresight to see what was needed, and made the sustained effort necessary to achieve it. This is followed in Chapter 3 by an account of the developments in objectives, institutions and methods which made possible the rapid expansion of activities which followed. The reconnaissance phase is covered region by region in Chapters 4-11. One country, Malawi, is described in more detail in Chapter 8, as a case study based on personal experience. In Chapter 12 the perspective is widened, with summaries of surveys of climate, geology, vegetation, and land use.

The phase of consolidation lies mainly outside the objectives and time frame of this account, but I have indicated who were the first local surveyors and when they took over from expatriates. The existence of soil survey departments or units in almost every developing country indicates recognition of the continuing need for their work by national governments. Visiting these countries, it is common to find that the surveys produced during the reconnaissance phase are still being taken as a framework for local work. In land resource survey, the legacy of the Colonial period has not been lost.

Interspersed with the accounts of surveys are biographical sketches of the people who made them. Some are placed in the countries with which they were most associated, but a few widely-travelled individuals are taken into a separate chapter. Many of the correspondents who supplied material added unusual, instructive or bizarre happenings in the field — some are unwise actions by newly-arrived expatriates ignorant of local conditions. I have included a selection, mostly to be found at the end of country accounts under the subheading 'Recollections'.

Chapter 13 takes up the story of soil erosion and conservation, how earlier attempts at conservation which met with only limited success were transformed into an approach first called land husbandry and subsequently conservation agriculture.

Chapters 14 and 15 are retrospective, reviewing work from the beginnings of the pioneering phase to the transition from surveys by expatriates to ongoing work by local staff. In Chapter 14 the lives and work of the surveyors are reviewed, including the subsequent careers of those who left field survey. Chapter 15 summarizes what was achieved, including mistakes that were made and where things could have been done better. It also draws attention to the fundamental fact that agriculture and all other kinds of rural land use are dependent on natural resources, and thus on their management and conservation. This might seem obvious, yet at the present day, with its emphasis on social aspects and participation by farmers, it is in danger of being lost. The continuing role of land resource scientists in development and advisory work has become still more important, now that pressures on land are much greater, and the need for sustainable use so widely recognized. The final chapter reviews needs, over and above survey, for today's soil scientists to study soils in the field.

2

THE PIONEERS

Explorers and travellers

FARMERS WERE THE FIRST soil surveyors. Those who were shifting cultivators knew where and when to clear and plant, and used indicator plants (as we would now call them) to tell them where not to do so. Those who were settled on one piece of land knew it like the back of their hands, far more thoroughly than could be ascertained by the most detailed soil survey. Besides soil distribution, they had an intimate knowledge of applied soil science, such as where to plant the more fertility-demanding crops and where to put those which could tolerate poor land.

The first extensive knowledge of the tropical land resources to reach European attention came from the accounts of early explorers in Africa, seeking particularly to shed light on the larger lakes and the source of the Nile, and from subsequent scientific travellers. Whilst giving greatest attention on the natives (the word is non-pejorative) whom they encountered, they described the topography and vegetation, sometimes because of the obstacles which these presented to their progress. Observations on agriculture and, less commonly, on soils are to be found. David Livingstone, besides his strong personal motivation of missionary zeal, writes:

> The main object of the Zambezi Expedition, as our instructions from Her Majesty's Government explicitly stated, was to extend the knowledge already attained of the geography and mineral and agricultural resources of Eastern and Central Africa ... and to endeavour to engage the inhabitants to apply themselves to industrial pursuits and the cultivation of their lands ... It was hoped that by encouraging the natives to occupy themselves in the develop-

ment of the resources of the country, a considerable advance might be made towards the extinction of the slave trade.

In 1865, crossing country that was to become part of Malawi, he observes, 'We saw large tracts of this rich brackish soil both in the Shire and Zambezi valleys, and hence, probably, sea-island cotton would do well.' The following year, in the midst of a description of the horrors of the slave trade, he writes, 'Immense quantities of wood are cut down, collected in heaps, and burned to manure the land, but this does not prevent the country having an appearance of forest.'

Sir Harry Johnston, after an early career as a Colonial governor, devoted his polymath abilities and immense energy to scientific exploration. The quantity of information in his *The Uganda Protectorate* (1902) is phenomenal: two volumes totalling 1016 pages, on the landscape, history, commercial prospects, meteorology, geology, botany, zoology, anthropology and languages, all compiled whilst he was supposed to be governing the country. In *British Central Africa*[1] (1897) his description of shifting cultivation parallels that of Livingstone, but he was possibly the earliest European to deplore this practice:

> The natives make clearings for their plantations. They cut down the trees, leave them to dry and then set fire to them and sow their crops amongst the fertilizing ashes. The forest is replaced by grass or scrubby trees which can resist the annual scorching of the bush fires, which play a considerable part in the disforesting of the country, so that year by year it diminishes in area to extinction.

These are sentiments which would meet with approval today. His most eloquent passage is a tirade about insects: tsetse-flies, mosquitoes, jiggers, sand-flies, ants, locusts, hornets, and cockroaches. Insects are:

> That class [of life] which seems to have been created for an almost wholly evil purpose … From the point of view of man and most other mammals insects are the one class amongst their fellow creatures which are uniformly hostile and noxious … Birds have nobly devoted themselves to keeping down insects, and for this reason deserve the gratitude and support of humanity … This

---

[1]   Which became Nyasaland and later Malawi.

declamation may appear strained in its tenour, but a prolonged residence in any part of Africa produces in one's mind a sweeping hatred of the insect race.

Quick to recognise these needs, the Department of Agriculture appointed its first entomologist in 1911.

For soils and their agricultural potential, the most striking exploratory account is a journey undertaken by Lord Lugard (1858-1945) in what was then Uganda, covering also much of modern Kenya. His account is in the form of a map, along the traverse routes of which are written hundreds of notes in tiny red writing. There are frequent mentions of soils, together with their estimated richness and actual or potential agricultural use:

> Soil black but dried up, grass very coarse useless for fodder ... Grass plain, soil fair but parched, stunted *acacia* and *euphorbia* ... Interminable cultivation and villages, population very dense, very timid and peaceful ... soil perfection, *shambas* well kept ... Valleys a rich black swamp soil, hills of red mail and quartz and shale gravel, iron-ore slag ... Hills with low angle of slope, covered with grass knolls of marl, iron-ore slag and gravel [this is murram or laterite], valleys swampy, suitable for rice culture.

Early scientific accounts

Among scientific accounts of tropical soils, priority should be given to the description of laterite in Malabar, India, used as a building material, by Francis Buchanan in 1807. For agricultural soils, an early account is *The Soils and Agriculture of Penang* (an island off Malaysia) by Captain Low in 1836.

There is a report of 1873, *Remarks on the chemical analyses of samples of soil from Bermuda*, written by His Excellency Major General J.H. Lefroy, R.A., F.R.S., Governor, President and Commander-in-Chief. One might not expect someone with such duties to be strong on agriculture but many people at that time were polymaths, and he is well aware of the findings of Baron von Liebig, the founder of agricultural chemistry. 'These remarks make no claim to originality ... The writer has aimed at nothing beyond the application of established principles to local conditions ... in a form intelligible to unscientific readers and to practical Agriculturists' — would that later surveyors had more often followed this precept. The chemical properties of Bermuda 'red soils' and 'white soils' are contrasted and compared with the composition of British [Rothamsted?] soils, concluding that the red soils should be able to grow excellent wheat. 'Soils are classed as calcareous which contain as much as 20% of lime' [compare the modern approach to quantitatively-based soil classification]. 'A great deal more stock should be kept, because stable manure has in it *every fertilizing quality for all soils*' [italics in the original]. The report ends on a practical matter, the design of a patent earth closet to improve hygiene at St George's barracks.

Many other early accounts consist basically of soil sampling and analysis. In 1879 the Ceylon Coffee Planters' Association arranged for an agricultural chemist, John Hughes, to tour their estates. He issued instructions to managers to take samples, including:

- Sample to a depth of 1 foot [30 cm] from at least six places, and mix together.

- For comparison, send a sample of good coffee soil.

- Accompany the samples with information on ten items, including elevation, rainfall and drainage, number of years under coffee, types and amounts of manures, average crop returns, and 'any relevant information on the past history and present condition of the soil.'

Although not a survey in the modern sense, this shows an appreciation of soil variability, soil effects on crop management, and the possibility of soil degradation. A similar exercise for forest soils of Sri Lanka was conducted by Alexander Bruce in 1923, although these are purely textural and chemical analyses, with neither soil descriptions nor horizon differentiation.

V.H. Kirkham was the first to use the term 'soil survey' in a leaflet on methods of sampling issued by the Kenya Department of Agriculture in 1913, and the following year he published an account of samples taken from Guyana. C. Harold Wright, Agricultural Chemist, Fiji, 1914-1922, published a *Report on the soils* as a Departmental Bulletin in 1916. In Burma, soils of experimental farms were described in 1916, and in 1927 a study was made of soil factors causing the death of teak in plantations. The most noteworthy work, however, was by James Charlton, Principal of the Agricultural College in Mandalay, who made two detailed soil surveys in 1932, and three years later wrote an account of the sugar cane soils of Burma.

Personal visits to national survey offices, trawling through early bulletins and unpublished reports where they survive would doubtless turn up other early studies. None of the early work is 'pure' pedology, or soil genesis, of the kind that was being developed in Russia and elsewhere at the time. It is directed at the effects of soils, favourable and adverse, on plant growth and crop production.

Between the World Wars: the pioneer surveyors

The accounts so far noted were useful early efforts, but not surveys in the modern sense. The pioneers of soil survey were the following:

- Frederick Hardy, College Lecturer, Barbados, 1911, Soil Chemist, Imperial College of Tropical Agriculture, Trinidad, 1920; earliest published surveys of Dominica and Montserrat, 1922.

- Arthur Hornby, Agricultural Chemist, Malawi, 1921; published Soil Survey Bulletins from 1921 onwards, and a *Soil Map of Central Nyasaland* in 1938.

- Frederick Martin, Senior Assistant Government Chemist, Sudan, 1919, and Agricultural Chemist, Sierra Leone, 1924; began a national soil survey in 1924 and published a *Soil Map of Sierra Leone* in 1932, the first national map from the Colonies other than those of small island states.

- A.W.R. Joachim, Agricultural Chemist, Sri Lanka, 1925; from 1935 conducted local surveys of the Dry Zone, and compiled a national soil map in 1940.

- Geoffrey Milne, Soil Chemist, Tanzania, 1928; published reconnaissance accounts and map of the soils of East Africa, 1935.

- Colin Trapnell, Ecologist, Zambia, 1932; published the first part of an Ecological Survey of the country in 1937.

There were other soil scientists active between the wars, as noted in the accounts of individual countries. The above six, however, may be granted special status, as the first major land resource surveyors in British overseas territories. Readers may decide whether the laureate for the earliest should be awarded on grounds of first appointment, first evidence of survey activity, or first published soil map.

Frederick Hardy (1889-1977)

Frederick Hardy had an active research career spanning over fifty years; additionally his influence as a teacher was profound. Almost all recruits to the Colonial Agricultural Service from the 1920s onwards spent a year at the Imperial College of Tropical Agriculture, Trinidad. Hardy was its founding Professor of Soil Science, and on the opening day of the College in 1921, at 8.00 a.m. precisely, gave the first lecture. He is described as being of the old school of gentlemanly decorum, but with a reticent manner, and was reluctant to be photographed.

Original staff of the Imperial College of Tropical Agriculture,
Trinidad, 1922; Frederick Hardy standing left

Between the wars, all British candidates intended for the Colonial Service would come into contact with Hardy's down-to-earth, agriculturally orient-ated, approach to soil science. He could hold the attention of a class of postgraduate students who had already spent a year at Cambridge, and thought they were the cat's whiskers. He would walk into the lecture room, turn to the blackboard and write, 'It all depends', then wait in silence as, not knowing what to expect, would the students. Then, fully in command, he could begin to set out the way in which each set of environmental condi-tions needed to be managed differently. His lectures were geared to an ecological philosophy, based on the mantra, 'The soil, the plant and the atmosphere are three components of a single system.'

Over much of his working life, Hardy spent the vacations conducting surveys of the various islands of the West Indies. His approach was first to find out which crop or crops the export economy of the island was dependent upon, mostly cocoa, bananas, sugar cane or coconuts. If the dominant crop was, say, cocoa, he would arrange a tour of the main estates. 'Take me to where your best cocoa grows' he would say to the manager, 'and take me to places where it is not doing so well.' He would cause pits to be dug on both 'good' and 'bad' sites, and then follow the normal soil survey procedure: field profile description, with particular attention to drainage and soil physical conditions, and sampling for analysis. By this means, the agricultural performance of the soils, with the respect to the specific crop, was associated with the soil characteristics right from the start.

This system may be called the Hardy method of soil survey. It was followed through ten surveys over 1922-1936 and three more in 1947; eleven were of islands in the Lesser Antilles, together with Jamaica and Belize. They were published as *Studies in West Indian Soils I-XIII*, and are generally referred to as the 'Grey Books' (to distinguish them from a post-war series of West Indian surveys of a different nature, the 'Green Books'). Fieldwork for each survey occupied one to three weeks.

The reports follow a standard pattern, illustrated by No. V, *The sugar-cane soils of Antigua*, from which a number of key points emerge. The first comes in the opening paragraph of Acknowledgements: 'The authors wish to thank ... those planters who ... furnished information regarding yields and field experience.' Thus, management experience was known before any soil was described. The reports begin with the standard elements of soil surveys: descriptions of topography, climate and geology, methods of analysis, and discussion of laboratory data. Then comes 'Agricultural classi-fication of Antiguan sugar-cane soils', of which there are four classes: Non-calcareous and Calcareous, each divided into Non-saline and Saline. In most of the other surveys the number of soil types is larger, and for the most part geologically based. The next section is on 'Agricultural relationships', identifying, 'the several factors which appear to be responsible for dimin-ished fertility,' followed by 'Recommendations for Improvement': to water supply, air supply [drainage], nutrient supply, and mitigation of 'harmful factors'. The map (scale 1:140 000) is rudimentary, consisting of the geolo-gical map on which are marked sites of soil pits.

In most of the Grey Books there is no soil map as such; in No. VI, Jamaica, it is based on a geological map of 1898. Hardy would have been horrified, however, if it were suggested that these were not true soil surveys. What mattered to him was that the ways in which soil properties affected desirable land-management practices had been established: drainage on one kind of soil, fertilizer on another, whilst some soils were best avoided for cultivation altogether. He knew full well that the distribution of soil types was so complex that mapping, to locate areas needing different treatments, was best left to the estate managers.

These discussions of agricultural management are applicable only to the crops already being grown. If it were proposed to introduce a new crop, say to make use of the more saline soils of a country, it would be necessary firstly, to draw on knowledge of that crop derived from elsewhere, and secondly, to map the degrees of salinity before planting. Hardy would hold that this was the right order in which to do things. Others believed that soil surveys should be general-purpose, mapping the distribution of soil types irrespective of what they might be used for; special-purpose surveys, for specific crops, management practices, or other interpretations, could follow later. By this view, surveys focused only on specific crops were liable to become useless in the longer term, as market conditions and land use changed.

A long-forgotten debate on this question took place in print in 1929-30. There had been a soil survey in Java by a Dutch surveyor, Arrhenius, which identified acidity as the main soil factor affecting sugar cane yields and therefore focused on this property, somewhat as in the Hardy method. This was in contrast to general-purpose surveys, which mapped soil series irrespective of their potential, as promoted by the influential US Soil Survey. A review of Arrhenius' survey by Willcox set off a debate in an obscure journal, *Facts About Sugar*. It was argued on one side that, 'Conventional soil maps ... are futilities,' and the only factors worth taking the trouble to locate on a map were those which affect crop yields. This was opposed by the view that, 'Soil science must deal with the soil complex as a whole'. Hardy was of the former opinion:

> Soil surveys of the type that were largely in vogue up to the early
> years of the present century have not proved as valuable as was at
> one time hoped. Many farmers have even maintained that they

could classify soils better than the specialist soil-surveyor merely by observing plant-growth … and by 'walking the land'.

The type 'in vogue' would now be called general-purpose surveys. Hardy advocated taking pairs of sites, 'good' and 'bad' according to the opinions of planters, and comparing their soil properties. As early as 1930, however, he forcefully advocated that, 'An accurate knowledge of the potentialities of the soil must always form the foundations of any schemes of agricultural development and land utilization.'

Hardy's view lost ground during the post-war era of development projects and land settlement schemes. It was only later that soil scientists began to realize that their maps were not actually being used, leading to a disillusion among governments and even the discontinuation of some national surveys; the pendulum swung back towards linking crop performance data with soil types.

After retiring from his post in Trinidad in 1955, Hardy joined the Inter--American Institute for Agricultural Science in Turrialba, Costa Rica. There he consolidated his position as the leading expert on the soils and environ-ment of cocoa. 'Retiring' again in 1966, he went back to live in Trinidad where he continued to go into his office on most days, even when nearly blind. Nazeer Ahmad, a former student of his, and his successor as Professor of Soil Science, writes, 'The study of pedology was his hobby'. He died in 1977 at the age of eighty-eight. Truly, Fred Hardy has been called the doyen of British tropical soil scientists.

A word should be added on G. Rodrigues, Hardy's analyst for the pre-war surveys. Rodrigues was white, originated from the Canary Islands, and spoke unaccented English. There was a term in soil physics called the 'sticky point', the water content at which clays first became sticky. Hardy so often promulgated this that he was sometimes known as 'Sticky-point Hardy' but the concept was at least equally due to Rodrigues.

Frederick Hardy established soil survey, with a strong link to improved crop production, as an essential basis for agricultural improvement in the tropics. Through his teaching over many years he had a widespread impact, first on those who were to become district agricultural officers, making them aware of the highly varied nature of environmental conditions; and secondly on

Colonial Office policy, leading to the subsequent development and funding of surveys. He was the foremost teacher of tropical soil science.

Arthur Hornby (1893-1955)

Appointed Agricultural Chemist, Malawi in 1921, Arthur Hornby remained there throughout his working career, continuing to live in that beautiful country during his years of retirement. His achievements are due to the foresight of two men: the first was the Director of Agriculture, E.J. Wortley, who very shortly after appointment allotted part of his miniscule budget to a post of agricultural chemist; the second was Hornby himself. Appointed to conduct soil analyses for fertilizer advisory purposes, soon after his return from leave in 1921 he decided to establish a national Soil Survey, staffed by himself, and published his first *Soil Survey Bulletin* in 1925.

There were fellow enthusiasts for landscape in the country at the time: J.B. Clements, Chief Forest Officer, and Frank Dixey, Government Geologist, later to become Director of the Colonial Geological Survey. Together they travelled through the country, and were struck by the erosion which followed clearance of vegetation on sloping land. Their views were set out in a locally published bulletin of 1924, *The destruction of vegetation and its relation to climate, water supply, and soil fertility*, each co-author writing a section. 'The enormous loss in soil fertility on the cultivated and cleared slopes of the Shire Highlands ... due to the washing away of the top-soil, is apparent to every observant person.' This was when the population was about 1.5 million, less than a tenth of its level today. Hornby calls for conservation of soil organic matter and fine particles, setting out means for doing so, by decreasing the velocity of surface run-off and increasing the absorptive capacity of the soil. One of six means for reducing run-off is 'Contour hedges ... Hedges of leguminous and other plants are often closely planted along the edge of terraces', a proposal which anticipates the use of this technique in agroforestry by some sixty years. Dixey notes that deforestation may lower the level of the ground water, and Clements lists eight beneficial effects of forests, which would now be called their environmental services. He set out his views on erosion on conservation in 1934, and publicized them through articles in the *Nyasaland Times*.

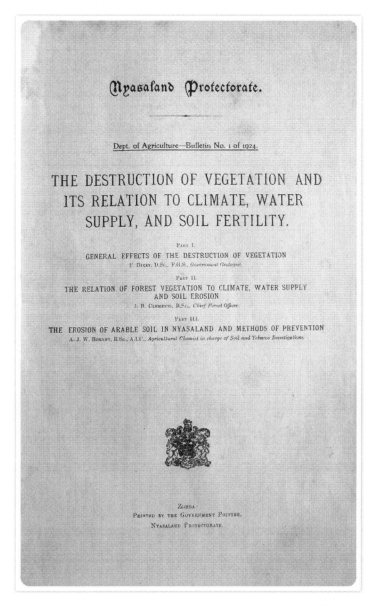

An early account of the environmental damage caused by
vegetation clearance, in the somewhat austere format of
Government publications of the day

In his early years he set about soil reconnaissance studies, issuing a series of
Soil Survey Bulletins during 1925-1930. He then widened his activities into
agricultural surveys of the greater part of the country, calling upon the
assistance of district agricultural officers, the entomologist, and the chief

veterinary officer. He was the more readily able to secure their cooperation having become Assistant Director of Agriculture from 1931. A junior officer who supplied information for Likoma Island, in the middle of Lake Malawi (then Lake Nyasa) was R.W. (Dick) Kettlewell who was to become Director of Agriculture, and later Minister of Natural Resources at the time of Malawi's independence. Hornby's ensuing reports were on Central Malawi in 1935 and Northern Malawi in 1938, the former in mimeographed typescript. The bulk of the text is on a district basis, covering for each, climate, soils, vegetation, and agriculture. He then wrote summary sections on soil types and fertility, erosion, grasses, and population. There is a 'Diagram of Agricultural Division of Nyasaland', a two-axis chart setting rainfall against elevation and fitting the vegetation types onto this.

His crowning achievement was a *Soil Map of Central Nyasaland* (1938), the culmination of seventeen years of survey. Given that the topographic base is somewhat primitive, this is an astonishing map for its time. The major limitation, as with all pre-war surveys, was the absence of air photographs. The map was put together by interpolation from an intricate network of road and track traverses. Despite these problems, the soil types and their distribution show a substantial accord with modern maps. It is accompanied by a map called 'Land Utilisation' but which would nowadays be termed 'Land Use Potential'. It shows 'Main areas suitable for economic crops' such as cotton, tobacco, sunflower and groundnuts, all of which were, in post-war years, to become export crops of the country.

A striking feature of Hornby's career was his independence of action. A chemist by training, he worked largely without contact with the academic world apart from getting himself sent on a study tour with the United States Soil Conservation Service in 1925. The natural landscape itself was Horn-by's teacher, observing the complexity of the environment in the field, and working out relationships between its factors. He continued to live in Malawi on retirement.

Arthur Hornby worked in Malawi throughout the inter-war period, his work comprising soil conservation as well as survey. Although not widely known outside the country, by moving to Director level he ensured that attention would continue to be paid to environmental and conservation aspects of agriculture, and he contributed to awareness of survey needs at the level of the Colonial Service as a whole.

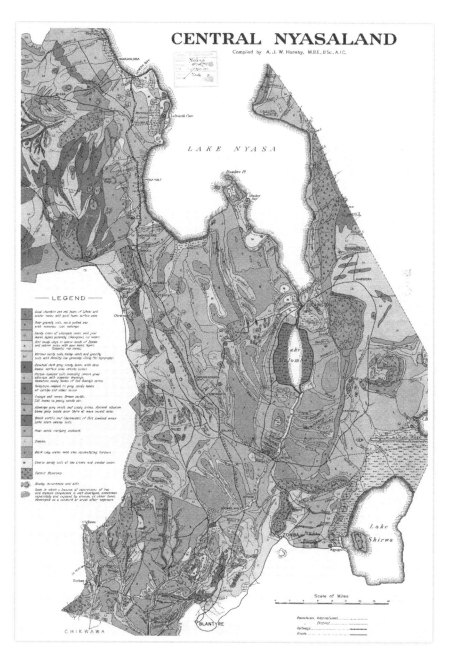

Arthur Hornby's Soil Map of Central Nyasaland (1935), covering
what are now parts of Southern and Central Regions, Malawi.
The original is in brilliant colours at scale 1:350 000

Frederick Martin (1891 – c.1964)

Frederick Martin followed the same career pattern as Hornby, although his period of active survey was shorter. By the age of twenty-eight he had fitted in a Cambridge natural sciences degree, Diploma of Agriculture, a PhD, and five years of war service in France. He became Senior Assistant Government Chemist, Sudan, in 1919, and was then appointed Agricultural Chemist, Sierra Leone, 1924, finding himself in charge of the Division of Research, Lands and Forests Department. He began the work which he was doubtless expected to do, soil sampling and analysis, and by the end of 1925 had completed this routine work for 639 profile pits. Early on, however, he realised the need for a mapping basis to the numerous sampling sites. Hence, 'A soil survey of Sierra Leone was started towards the close of 1924, the object in view being to endeavour to examine and classify the various types of land in order to obtain an idea of the agricultural possibilities of the country', an admirably concise and comprehensive statement of objectives, not least in referring to 'land' rather than just soil.

The survey proceeded rapidly, possible if someone is not only dedicated to the work but also allowed to get on without the interruption of administrative duties. By July 1926 he had produced a *Report on the survey of the soils of the Colony and Protectorate of Sierra Leone*, a preliminary report which is largely a collation of analytical data. A soil map of the whole country was completed in 1932. It is at 1:1 000 000, in black and white, hand-drawn at the Survey Department in Freetown. Five soil types are shown, with simple descriptive names such as 'Coarse sandy soils' and, the most extensive, 'Laterite and lateritic gravel'. Lacking a means of mapping boundaries, it plots symbols (crosses, lines, etc.) but not boundaries on the map, leaving a margin for local discretion. Although making the transition from analyst to surveyor, Martin's soil profile descriptions follow ad hoc methods, and he lacked an appreciation of the geographical relationships between soils and landforms.

He did, however, display independence of thought when it came to questions of laterite. Finding an early definition by Fermor that 'laterite' referred to soils containing 90-100% hydrated oxides of iron and aluminium, whilst 'lateritic soils' contained 25-90%, Martin remarked that, 'No samples of reputed laterite soil have yet been found … which will conform to Fermor's definition.' He also noted that geologists when speaking of laterite refer to the hard concretionary material present in some soils, not to a soil type as a whole. So Martin put forward his own definitions: if the silica:alumina ratio of the clay fraction is below 1.33 the soil is a laterite, if it is between 1.33 and 2.00 it is lateritic. These definitions were not subsequently adopted, nor did they remove the confusion which for long reigned over laterite, but in applying quantitative criteria to soil classes he anticipated the approach which came to dominate soil classification some thirty years later.

Martin was active in the wider encouragement of soil survey, speaking at a West African Agricultural Conference in 1927, and the same year encouraging standardization of analytical methods through the Soils and Fertilizers Committee of the Imperial Agricultural Research Conference. He subsequently moved out of the scientific field, to become Assistant Director of Agriculture in 1930 and Director 1934-1943.

Frederick Martin was an isolated figure, not previously recognized as a pioneer. He perceived the need for survey as a basis for soil chemical studies and, largely single-handedly, completed the survey of Sierra Leone, the first national soil map. Moving to Director level, he became active in the planning of soil surveys for the Colonial Empire as a whole.

A.W.R. Joachim (1898-1979)

Anyone perusing the Colonial staff lists for Sri Lanka and coming across the name 'Anian Walter Richard Joachim', with a London BSc and a Cambridge Diploma in Agriculture, might be forgiven for envisaging a keen young Englishman, eager to make his mark in the Colonies. This is far from being the case. Joachim was a Sri Lankan national, belonging to a group known as the Burghers, descendants of Portuguese and Dutch settlers in the seventeenth century. As such, he was to become the earliest soil surveyor of local origin, his appointment just predating that of Mohammed El A'al in Sudan.

At St Benedict's College, Colombo, Joachim was an outstanding student, excelling in mathematics and chemistry, and ranking first in the country. It is said that because of his padre-like appearance his mentor, Brother James, had visions of attracting him to a religious life; but when in 1919 he was awarded a British Government scholarship Sir Marcus Fernando, the leading physician of the day, advised him to take agricultural chemistry.

Returning to Sri Lanka, he was appointed Agricultural Chemist in 1925, finding himself among agricultural specialist staff who were largely British. One can imagine there was a certain camaraderie, for looking through staff lists it is noticeable that, once posted to Sri Lanka, officers showed little desire to leave. For the first five years Joachim conducted studies on animal feedstuffs and green manuring.

A. W. R. Joachim

From 1930 onwards he embarked, largely single-handedly, on a series of soil surveys in the then uncleared jungles of the country's Dry Zone. The results were presented in a series of fifteen papers in the journal *Tropical Agriculturalist*. In these days of rapid communication it is hard to put oneself in the situation of an isolated surveyor, and to know where the links with developments in land resource studies were made. Attendance at the Third International Congress of Soil Science at Oxford in 1936 would certainly have made a contribution, for at international meetings before the Second World

War one could talk to everyone of importance. Travelling round the country on local surveys, Joachim devised a practical, descriptive soil classification, and also acquired a geographical eye for the relations of soils to landscapes. By 1940 he had drawn up a national soil map, although owing to the exigencies of the war years it was not presented until 1945. It is on a very small scale and, like Martin's map of Sierra Leone, shows soil types by line-shading but without precise boundaries.

Coupled with his distinguished research record, Joachim possessed an administrative ability which impressed his superiors. In 1947 he was appointed Acting Deputy Director of Agriculture, and in 1950 succeeded Sir Frank Stockdale as Director, among the earliest local staff in the Colonial Service, and one of the few soil scientists, to be appointed to this position. Remaining in this post until 1956, 'retirement' was far from the end of his activities, becoming Director of the Tea Research Institute, Chairman of the Land Utilization Committee, and other government appointments. With the combination of long service and high office, it is not surprising that honours should come his way: MBE, OBE, and in 1953 the Coronation Medal (along with members of the first successful Mount Everest expedition). Sri Lanka is unique among British colonies in having its land resource survey between the wars led by a scientist of local origin.

Geoffrey Milne (1898-1942)

With Geoffrey Milne we come to one of the giants in field observation of soils and landscapes. He is remembered for three achievements: the concept of the soil catena, which was to be widely used as a basis for mapping at reconnaissance level; his 1936 soil map covering the four territories of British East Africa; and for first observing the relationships between landforms, soils and vegetation that are characteristic of a wide area of East and Southern Africa.

Milne was born in 1898, the second of three sons of Sydney Arthur Milne, a school headmaster in Hull, and won a scholarship to Leeds University; a brother became Rouse Ball Professor of Mathematics at Oxford. His achievements are the more remarkable in that as a field scientist he was largely self-taught, being trained as a chemist and starting his career as a lecturer in agricultural chemistry. As to how he acquired an interest in the field study of soils, there are clues to be found in his first two appointments, at Aberdeen (1921-1926), location of the Macaulay Institute of Soil Research, and at Leeds (1926-1928). Leeds possessed a notable Department of Agriculture, but in addition he had the good fortune to meet and marry Kathleen Morgan, a geography graduate who had studied under H.J. Fleure at Aberystwyth. Academic geography at that time was not scientifically sophisticated but it fostered a tradition of field observation. Kathleen extended his interests beyond the somewhat self-contained ambit of British agriculture into the wider world, and also taught him that knowledge of landscapes was not a matter of soil sampling and laboratory analysis but came from travelling and recording what you saw, particularly the relation-ships between the different elements of the environment — geology, land-forms, water, soils and vegetation.

Geoffrey Milne

With the title of Soil Chemist, he was given charge of soils research at the East African Agricultural Research Station, Amani, Tanzania, one of the transnational institutions which played such a valuable part in Colonial agri-cultural research. There is an unusually good record of Milne's methods, since after his untimely death Kathleen not only wrote detailed accounts of his life but collected his papers, now preserved in Rhodes House Library, Oxford.

The foundation of his work lay in field traverses. Ten field notebooks exist, covering 1937-1941. They are written in tiny but legible writing, mainly in pencil, in the small (15 x 10 cm) government notebooks of the time. One is annotated, 'After taking John for his first term at St Andrew's school, Turi.'

Two Colonial Service colleagues also gave him guidance. Clement Gillman had the knowledge of landscapes that comes from the practice of geologists of spending over half their time in the field. Gillman was later to edit Milne's published reconnaissance traverse. B.D. Burtt, Ecologist, first went out to East Africa in 1925. He set about plant taxonomy, the identification of species, a task which is vital as a basis for fieldwork, although one which to outsiders does not appear exciting. Burtt worked systematically but slowly, and as his Director, Frank Swynnerton, wrote, 'At the time of his premature death [in 1938] he had not synthesized his knowledge.' Burtt's work was put together in a posthumous publication based on his departmental reports, the essence of which was a thirteen-page table listing some 360 tree, shrub and grass species with their habitat and regions in which they are found. It includes the major early description of *Brachystegia-Isoberlinia* tree and shrub savanna, generally known by the Swahili term *miombo*, the most widespread vegetation community over a wide area stretching from Tanzania to Zimbabwe and Mozambique.

Milne's three-month soil reconnaissance journey through Tanzania, 1935-1936, was subsequently transcribed and published in 1947. It is one of the finest examples of geographical field observation in existence, acute in the perception of environmental relationships, and so clearly set down that Gillman, in an introduction to the published version, writes that 'his fluent and expressive style requires no editing'. From Amani, inland from Tanga in the foothills of the Usambara Mountains, Milne travelled south-west to Dodoma, followed the railway west to Tabora, then went north to Shinyanga and Mwanza on Lake Victoria, resulting in a transect through all the main ecological zones of the country. Did he travel by vehicle or on foot? There is no indication, but in the contemporary record by Burtt, one photograph shows a small pickup truck, another a line of eighteen head-porters. Milne's objective was:

> ... to look at the relationships of the principal soil types to each other and to the various natural factors concerned in soil formation

... It was found that soil and vegetation zones are so linked that neither can be fairly discussed without the other.

Besides *miombo*, an extraordinarily wide range of features characteristic of the region are described, such as pisolithic murram [nodular laterite or ironstone], concentric weathering in diorite boulders, soil hard-pans exposed by erosion, the accumulation of calcium carbonate in termite mounds, and the *Itigi* thicket, the monotonous, uniform, scrubby vegetation found in the centre of Tanzania. The soil types Milne uses are descriptive: skeletal soils, red sandy soils, red soils with a murram horizon, calcareous grey sandy soils. There is an extremely widespread class of 'plateau soils', of low fertility although cultivable, and which modern systems of soil classification fail to recognize adequately; Milne rightly observes that these are 'in the main to be regarded as belonging to a past stage in the drainage conditions of the area.' Anticipating by some fifty years the modern practice of drawing upon indigenous knowledge, he gives the Sukuma names, e.g. *kikungo* for red-brown loam, *mbambasi* for hard-pan soils, *mashishewe* or *mashalova* for murram or concretionary subsoil ironstone.

No one crossing the plains of Africa can miss the grassy valley floors known in Tanzania as *mbugas* and further south as *dambos*. They are seasonally waterlogged and were formerly used mainly for grazing although under population pressure, cultivation extends down the valley sides into their margins. Milne recognized that although difficult to till, the black clay soils in these valley soils hold nutrients washed down the adjacent valley sides, both over the surface and within the soil. Cattle grazing in the *mbugas* are taken into *kholas* (cattle pens) in the villages at night, and Milne envisaged that this practice could be put to good use:

> With the aid of cattle and compost, there would be every prospect of bringing the limited areas of sandy *Brachystegia* soils into high production conditions, provided erosion losses were minimized by tie-ridging ... If the *mbuga* lands can by some such scheme be fully utilized and the produce [of grazing and *khola* manure] be directed to the sandy soils ... the purpose will gradually be achieved of *bringing fertility uphill* [my italics], counter-current to the natural tendencies.

Milne is most widely remembered and quoted for his concept of the soil catena. The catena is the succession of soils down a slope, from crest to valley floor — Milne's original definition was 'A regular repetition of soil profiles in association with a certain topography'. First formulated in his traverse reports, this reached a wider audience at the Third International Congress of Soil Science in 1935, at a time when congresses were small enough for participants to listen to all the papers. In the next year he followed a much to be applauded scientific tradition by presenting it in compact form in *Nature*. His account totals 700 words, comparable in length to the renowned paper in the same journal which first announced the structure of DNA.

What Milne had seen was that over large areas of relatively uniform topography, particularly the gently undulating plateaux that are so widespread in Africa, one meets the same soils types again and again between the watersheds and the valley floors. Often the soils on the crests are red, where the iron compounds are present in least hydrated forms; they become yellow on the slopes, and develop mottled lower horizons where the level of the wet season water table is reached. Driving a vehicle on an earth road there is a bump at each valley-floor margin as one crosses an outcrop of laterite, where the iron-rich water has come to the surface and hardened. Below this there may be a zone of sandy wash, followed by the black *mbuga* clay. There are many variants to catenas, the example in Milne's *Nature* paper being a rocky granite inselberg above a concave pediment.

Few concepts have proved more useful in soil survey than the catena, particularly in the tropics. When working at reconnaissance scales it is neither possible nor desirable to map out the distribution of every soil type. The catena can be taken as the basis for mapping, with one typical example described in detail. Agriculturalists working in a local area will rapidly pick out the boundaries between the component soils. Catenas were employed, for example, for the reconnaissance survey of Uganda. After the Second World War it was found that catenas could be readily identified on air photographs, and this formed the basis of the widely-used land systems approach.

Like other pre-war surveyors, Milne did not have the advantage of air photographs, and it is therefore the more astonishing that he was able to complete the *Provisional Soil Map of East Africa* at 1:2 000 000 scale, published in 1936 and covering the vast area of Kenya, Uganda and Tanzania. This must have been constructed by interpolation between his many traverses, coupled with reports from local agricultural officers. The map legend has colours for each soil type, based on descriptive classes, e.g. Red earths, non-laterised, on granites; Plains soils, calcareous; Plateau soils; Black and grey soils of bottom lands. On the map these are combined in vertical stripes representing catenas, for example the Buganda Catena.

There are also unshaded areas, some of which have the approximate soils written in the map as text, for example in Southern Tanzania, 'Great extent of monotonous country under dry forest (*miombo*) carrying mainly grey sandy soils (plateau soils) with red earths on footslopes of gneiss inselbergs.' A great merit of the map is that these white areas make it clear that further surveys are needed, helping to avoid the attitude so often found in modern times, 'Let us make use of existing data.'

At the time of Milne's work much of East Africa, particularly Tanzania, was occupied by natural vegetation with only scattered shifting cultivation. However, he was by no means seeking knowledge only for its own sake, and frequently commented on their agricultural potential. This was the justification for his strong advocacy of further surveys. At present, he observed, knowledge is acquired by individual agricultural officers and at best recorded, if at all, in scattered files:

> A survey is required in which distinction will be made between categories such as ... land suited for intensive, or for subsistence, agriculture; for productive grazing, rough grazing, wet-season grazing; for special management as a key area for water supply ... No such classification can be arrived at ... by haphazard study by individual departmental officers.

The last sentence of his Tanzania traverse reads:

> It is the pedologist's province to claim for soil survey that it shall be called upon for its proper contribution to the programme, in time for its results to play a part in the early framing of a rational land utilization policy.

In these words is found an early anticipation of the need for both land evaluation and land-use planning.

From the late 1930s he spent increasing time in advocacy of surveys, through publication and at conferences. In a 1940 memorandum, 'Soil survey in East Africa: some desirable developments', he wrote that:

> Mistakes can easily be made in conservation practice, upon soils that have not been thoroughly studied, as they can and have been in agricultural or pastoral practice. Soil surveys should precede the framing of conservation policy.

It was while visiting Nairobi, intending to take up a post of Scientific Secretary to the East African Industries Technical Advisory Committee, that he was taken ill with acute pancreatitis and died in mid-career at the age of forty-four.

Milne had three advantages for his work: the East African Agricultural Research Station gave him an institutional base with a mandate of six countries and adequate funding; in Clement Gillman, B.D. Burtt, and his wife, Kathleen, he had a group of knowledgeable advisers, sympathetic to his aims; and lastly, he had the once-only privilege of a geographical and scientific blank sheet, a sub-continental area with next to no systematic land resource studies. None of this detracts from his achievements, combining wide-ranging knowledge, sustained efforts in the field, powers of synthesis, and clear expression.

Geoffrey Milne was a man of extraordinary vision. Trained as a chemist, he acquired a perception of the natural landscape and the interactions between its parts, learning from colleagues, and from his geographer wife. Not least, an experience common to all field scientists, he learnt by travel and field observation of the land itself. His reports show an extraordinary appreciation of the nature of African landscapes and interactions between their elements, whilst his achievement in the completion of a soil map covering four countries is mind-blowing. In the concept of the soil catena, both as a functional unit and as a basis for mapping, he left a legacy widely applied in the post-war period.

Colin Trapnell (1907-2004)

Equal to Milne in his achievements — no purpose would be served in ranking one above the other — is Colin Trapnell. Trapnell had a lengthy career and continued to work long after retirement, to within a few years of his death in 2004. When I visited him in 2002 he was lucid, informative and amusing.

Uniquely among ecologists, he read Classical Greats at Oxford. At the same time, jointly with Max Nicholson, he founded the Oxford University Exploration Club and in 1928 participated in one of its first expeditions, to Greenland. On graduating, he wrote to the Colonial Office asking if there was a post available as an ecologist, an enquiry which might not have gone far had it not been supported by the father of British plant ecology, Professor Sir Arthur Tansley. At interview he was offered the post of Plant Taxonomist in Sri Lanka but was aware of the uninspiring (although essential) nature of such work, and declined. 'Just a moment, Sir' said the Secretary to the Board, 'I think we have a letter from an Oxford professor who recently visited Northern Rhodesia.' After drinking coffee whilst this was located, it was found to contain a recommendation for an ecological survey of the country.

The interview proceeded: 'Would you like this job, Trapnell?' 'Very much, Sir.' 'Do you have any ecological training?' 'No, Sir, only my expedition experience.' He was therefore sent on a course at the Royal College of Science, followed by a Colonial Agricultural Scholarship to visit research institutions, first in Britain and then in South Africa, with £30 expenses for accommodation and travel, including finding his way to Zambia. On arrival in Lusaka in January 1931 it transpired that in setting up this programme the Colonial Office had forgotten one detail, namely to inform the Department of Agriculture that he was coming, and arrangements were made for him to be taken on in a temporary appointment. Whilst on leave in Britain in 1934 the situation was clarified, and he was given a 'permanent and pensionable' post as Ecologist.

The ecological survey was based on the Department of Agriculture, in large part due to the personal interest of its director over eleven years, C.J. Lewin, who secured a grant from the Colonial Development Fund. It was carried out through an effort of dedication and sustained energy. Zambia is three times the size of the United Kingdom, and at the time was crossed only by an extremely sparse network of earth roads. Trapnell's observations were therefore made by means of extensive *ulendos*[1] or foot traverses, along roads where they existed, through the bush between villages elsewhere, and in the swampy centre of the country by boat. These traverses were conducted from 1933 to 1940. A photograph shows five field assistants and twenty-five carriers. Some field trips lasted up to three months, and every day he took copious field notes on vegetation, soils, and local agriculture, illustrated by sketches and numerous photographs. The country was divided for purposes of the survey into two parts, the western and southern area being surveyed 1933-1934, the eastern and northern area 1937-1940.

Colin Trapnell and Neil Clothier on safari in Zambia

Under the guidance of Lewin, good teamwork and a pooling of interests and abilities assisted the survey. Trapnell was joined in his early traverses by Neil Clothier, later (1950-1955) to become Chief Conservation Officer. Of him, Trapnell writes:

> In Neil I had an ideal colleague ... He combined a natural charm of
> manner and a self-deprecating turn of humour with an unconscious
> air of authority which ensured respect and cooperation from those
> he worked with. Neil was a practical man, versed in crop husbandry,

---

[1]   *Ulendo* is the Cinyanja (Cicewa) word applied in Zambia and Malawi to field
travel, corresponding to the Swahili *safari* used in East Africa.

who handled, also, the various problems of personnel and provisioning which arise on ulendo. He possessed the common touch, which brought a ready response in all his enquiries among cultivators and village headmen ... Throughout three years of strenuous ulendo we never quarrelled ... His influence, and all I had learnt from him, determined the whole future course of the survey as what has since become to be termed agro-ecology.

J.D. Martin surveyed the vegetation of Barotseland; he was killed on active service with the RAF in 1941. Others who joined Trapnell on traverses included U.J. Moffatt, the botanist Peter Greenway, and William Allan. It was out of this survey that Allan developed his ideas about how many people a given area of land can support, a concept taken up many years later in FAO's population-carrying capacity study. Trained as an entomologist, Allan was later to become Deputy Director of Agriculture.

Trapnell's traverse notes have been transcribed and published, occupying 1300 printed pages. They constitute a record of the country when it was sparsely settled which will remain of value for all times — comparison with the Domesday Book record of England when it was still partly under woodland is not inappropriate. Vital from the viewpoint of historical ecology is that the location of all the traverse routes is meticulously recorded, on maps and in the notes, so that scholars of the future can repeat them, knowing where and on which day the original records were made.

It is hard to convey the immense detail of this record in a short extract, and the following can only be arbitrarily chosen:

> **September 26th [1934] Kasempa** Through *Isoberlinia paniculata-Isoberlinia tomentosa (Brachystegia longifolia)* white sand-clay mixture, sand fine with no trace of Kalahari grains; broad shoulder of clayey fine sand. Route NE over head of small southeasterly dambo ... Local sorghum gardens and old village: bee cylinders. Small ironstone nodules exposed by dambo margin ...Some yellow argillaceous soil with <u>manga</u> *(Brachystegia hockii)* coming in instead of B. longifolia. [Similarly for a page] ... October 2nd The northern agricultural traditional Mwalukila [local Headman] maintains this yellow soil good ... He seems to mean that when burnt the trees mix with the soil and feed the crop. Large <u>chiteme</u> gardens cut beyond limit of cultivation on this poorer soil; some three times area cut to that

planted, in large blocks. Dull grey soils. Choose thickish bush for finger millet cultivation, if off anthill ... Say they will plant pumpkins and groundnuts and groundbeans in the finger millet patches the following year ... Through *Isoberlinia tomentosa* and small grey clay *Acacia pilispina* ... [three other species] chipya hollows ... Say they prefer this type for the sorghum and the ordinary *Isoberlinia paniculata* etc. for the finger millet. Make beer of finger millet or keep in store as emergency food. Also make ... sweet sorghum beer with munkoyo root. Get honey beer in January. Grey-buff lateritic-soiled sorghum gardens. Iron nodules, to *chipya* hollows and laterite bars beyond. [The lists of plant species are longer than quoted here.]

Although Trapnell does not use the term catena, he describes repeated topographical sequences and several of his diagrams illustrate soil-vegetation transects with their linked agricultural practices. There are frequent descriptions of the most widespread agricultural practice in Zambia, *chitemene*,[1] in which trees and shrubs are cleared from savanna woodland, piled up 1-2 m ,high on a smaller area in its centre and burnt, thus releasing potash, before planting a crop.

This vast mass of field observation was summarized in an admirably clear manner as two memoirs, appearing in 1937 and 1943, with a map which must have existed in draft at the time of these reports but did not reach publication until 1950. The *Vegetation-Soil Map of Northern Rhodesia* is in two sheets at a scale of 1:1 000 000. Indications of the sheer magnitude are the fact that it covers 12° degrees of latitude and 9° of longitude, and on this broad reconnaissance scale each sheet measures 60 x 90 cm. There are fifty-three mapping units, each consisting of a vegetation type on a group of soils, for example:

Eastern *Brachystegia-Isoberlinia* woodlands:
    On Plateau soils   P6
    On Escarpment hill soils   E1
Copaifera mopani woodlands:
    On brown Lower Valley soils   L1
    On grey alluvial clays   S1

---

[1] Trapnell uses *chiteme* but this term has entered scientific publications in the form *chitemene*.

The mapping units are coloured solid in bands along traverses and inter-vening areas of which he is reasonably confident, and in stripes of colour for infill areas surrounded by the same unit. With an honesty rarely found among surveyors there are areas, mainly steep dissected escarpment zones, left uncoloured. Anyone familiar with eastern and southern Africa will not be surprised to see that *Brachystegia* woodlands on 'plateau soils' are the most extensive unit.

To put vegetation as the primary basis for mapping, subdivided according to soil type, is unique to this survey, but was entirely appropriate to the sparsely populated nature of Zambia at that time. It was also the right basis for agricultural officers who would use the survey. They could recognize at once the vegetation types, after which the ad hoc descriptive soil types, like Kalahari sands or Plateau soils, presented no difficulty.

Whilst the survey was based on vegetation, the reports place emphasis on agriculture. Taking the two reports on North Western and North Eastern Zambia together, Part I, on vegetation and soils, occupies a quarter of the page space. Nearly half is devoted to detailed descriptions of some twenty indigenous agricultural systems, anticipating the later practice of farm system studies, whilst more than a quarter is concerned with potential for agricultural development by native farmers, the estate sector being outside the remit of the Department of Agriculture. The notion that Westerners could learn from local farmers only became a mantra of development orthodoxy only from the 1970s onwards, yet here it is stated clearly and in detail. In the Foreword to the 1937 memoir, the Director of Agriculture, C.J. Lewin, quotes from his own Departmental Bulletin of 1932:

> Recognition of the inherent soundness, under natural conditions, of native agricultural practice has only become general in recent years. Practices apparently contrary to the accepted principles of good farming usually prove on investigation to be the best possible in the circumstances under which the native cultivator works ... Under natural conditions, native cultivators have little or nothing to learn from the agricultural scientist, but their natural mode of life has been rudely interrupted ... Thus agricultural problems have arisen which were previously non-existent ... but *it behoves an agricultural department to investigate local practices with the utmost care before presuming to attempt to improve them* [my italics].

Apart from Hardy's studies of small West Indian islands and Martin's sketch map of Sierra Leone, the Ecological Survey of Zambia was the first to cover an entire country. The main users were district agricultural officers, probably giving more attention to the descriptions of farming systems, and development potential, than to the mapping of vegetation and soils on which they were based. One such, Charles Johnson (later to become Director of Agriculture, Malawi) lent me his copies, 'Somewhat the worse for wear I'm afraid, having been much used during my 24 years as an Agricultural Officer there.' Being carried out so early, the Survey was also available as a basis for planning at national level during the post-war era of development.

From 1958 onwards, an attempt was made to apply the ecological survey method to south-west Kenya. Martin Brunt, then Land Use Officer to the Directorate of Colonial Surveys, was seconded to assist Trapnell in the work. By this time, air-photograph interpretation had become the standard basis for field surveys, and was used to plot the boundaries of the ecological zones. Formal retirement made little difference to Trapnell's application to this work. Map sheets were distributed to district staff as they became available, although the final report was not published until 1987. It had much less impact than the Northern Rhodesian study, because mapping based on vegetation was intrinsically less valuable in populated areas.

Trapnell's long subsequent career, however, was not without achievements. In the immediate post-war years he was invited to Muguga, Kenya, headquarters of the East African Agriculture and Forestry Research Organization (EAAFRO) to train others in ecological survey, one of whom, Ian Langdale-Brown, became a noted botanist in Uganda. Another, George Jackson, spent nine years (1950-1959) as Ecologist in Malawi, and when I became Soil Surveyor there, taught me in the field how to link plant communities with soil types. Thus those of my students who subsequently went into soil survey can claim a teacher-pupil link with Trapnell.

In his eighties, supported by family money acquired through his brother's successful career as a banker and stockbroker, he established a Trapnell Fund for Environmental Field Research in Africa at Oxford University. The object of this was to integrate studies of climate, landforms, soils, vegetation, and the potential of land to support human populations. He continued to write papers until the age of ninety, particularly on the effects of repeated burning on natural vegetation, and in his last years assisted in the transcription of his Zambian field notes.

Trapnell's early interest in ecology was transformed by sustained dedication in the field. The sheer volume of his field observations, all of high quality, is possibly matched only by those of Hugh Brammer (Chapter 6). The Ecological Survey of Zambia stands as a unique achievement, appropriate to its time and place, even if this did not prove to be the method used subsequently. Through training and interacting with others he propagated awareness, first of ecology, and secondly of what should be leant from indigenous farmers.

Geoffrey Milne's starting point was soil science. Trapnell based his work on ecology. Both expanded their vision by prolonged field observation, taking in the whole complex environment, together with the farmers whose lives depended on it. If pressed to make a comparison, I would bracket Geoffrey Milne with Colin Trapnell as the greatest tropical field scientists of all time.

3

PREPARING THE GROUND

THE ACHIEVEMENTS OF THE pioneers were magnificent solo efforts, based on the limited financial support and technical methods available in their day. After the Second World War there was a rapid expansion in activity, leading to a golden age of land resource surveys from 1950 to 1980. The fundamental cause was a change in attitudes and objectives, a desire to improve the welfare of peoples in the Commonwealth, and thus to learn about the resources available for development. This led to the formation of institutions necessary to achieve these objectives, and was made possible thanks to the availability of new methods of survey.

Attitudes and objectives

A burst of interest and activity began in 1950. It might have started earlier but for the curtailment of activities caused by the Second World War, when staffing of the Colonial Service was minimal. The setback was not the six-year duration of the war but some ten years, since immediate post-war aid was focused upon the rehabilitation of Europe.

Three elements contributed to the expansion: changes in ideas and attitudes, the institutional development which accompanied these, and technical advances in survey methods. It is the changes in attitudes which are the most important. Once the attitudes, and hence the objectives, are clear, then the necessary institutional developments will follow.

One change arose directly out of wartime experience: a heightened concern for improved security in world food supplies. This was initially centred on restitution of food imports to Europe. Later, in the 1960s, came the realiza-tion that population increase in developing countries themselves would call for greatly increased food production if chronic hunger, and recurrent famine, were to be avoided.

Over and above food shortages, however, was the post-war move towards thinking in terms of development, the improvement of the welfare of poorer countries by means of investment. This was not entirely new. Improving the well-being of Colonial peoples in order to lead them to a position in which they could govern themselves had always been central to the objectives of the British Commonwealth. A Colonial Development Fund had existed in the 1930s, although it operated on what seems today an exiguous financial scale. But in the post-war period, development moved into the forefront of international policy, led by institutions set up by the United Nations and the World Bank.

Another incentive which had its origins before the war was concern over soil erosion. Local staff were aware of the adverse effects, in some cases irreversible, which this could have on production. In many countries of Africa and Asia, reports appeared, in some cases illustrated with graphic photographs. Lord Hailey's *African Survey* (1938), a massive review of problems arising in Africa south of the Sahara, contains a substantial chapter on erosion, written by Elspeth Huxley, which draws these accounts together. In a 1949 review of soil conservation in the British Empire, Harold Tempany has a section on 'Surveys as a means of providing essential information.' 'Soil maps ... by delineating areas possessing similar levels of inherent fertility and physical features can indicate cultural practices and methods of soil conservation suited to particular areas and crops.' Long before the invention of the modern terms 'land husbandry' and 'sustainability', agricultural officers were well aware of the need to combine improved production with conservation of the resource base on which this depends.

The concerns for food supplies, welfare of local populations, and conservation provided the incentives to learn more about the resources on which improvements in production would depend. Two conditions were needed to set in motion the expansion of land resource surveys: institutions and methods.

Institutions

The Colonial Service

Land resource surveys from the post-war years onwards until the times at which countries reached independence were conducted primarily by officers of the Colonial Service. There is a framework of four tiers: the Colonial Service as a whole, Departments of Agriculture, the Research branch of these, and within the latter the specialist staff in soils and ecology.

The first mention of a Colonial Service was in 1837, preceding the main era of Colonial expansion, the 'scramble for Africa'. Throughout the nineteenth century, this functioned through the services of individual territories, and most staff were administrative, legal or medical. A unified Colonial Service was not recognized until 1930, although an annual 'Colonial Office List' appeared from 1862. The Indian Civil Service (1858-1947) was always separate.

Professional staff grew in numbers in the early years of the twentieth century. In British tropical Africa by 1914 there were 1399 officers, of whom 538 were administrative, 258 medical, and the remainder in other professional branches: public works, surveys, education, railways, transport and marine, forestry, agriculture, and veterinary. From a land resources viewpoint, the first requirement was for topographic survey. Foresters slightly outnumbered agriculturalists. A rapid expansion took place after the First World War. In 1920 alone, new recruitment was 49 nine staff to agriculture and 33 to forestry, continuing on this scale until 1928, with 59 agriculturalists and 11 foresters. Staff numbers fell off in the Depression years of the early 1930s, then rose again before the Second World War. Even so, the Colonial Service remained on a very small scale: by 1936 there were 4400 officers stationed across 44 territories, of whom 800 were in the various branches of natural resources.

After the reduction in numbers during the Second World War, the passing of the Colonial Development and Welfare Act in 1945 led to a rapid expansion. This was accompanied by a change in emphasis from administrative to professional. Even by 1947 staff numbers had risen to 11 000, of whom 9000 were in the professional departments. The fact that intentions were followed by action in such a short space of time seems extraordinary in our present age. Recruitment proceeded apace, reaching a maximum of 1546 new officers in 1950. Peak years for the natural resource departments — Agriculture, Forestry, Geological Survey, Survey, Veterinary, and Meteorological Services — were 1948-1949.

In 1954 the Colonial Service was renamed Her Majesty's Overseas Civil Service (H.M.O.C.S.). The transition from British rule to independence began with India and Pakistan (1947), Sri Lanka (1948), and Sudan (1956). The main era was 1957-1966 when twenty-four countries were granted independence. A further twenty, mostly small island territories, reached independence 1967-1983.

In earlier years, each territory had its own separate Department of Agriculture (sometimes including Forestry). In 1935 these were combined with the creation of a Colonial Agricultural Service, but this change was largely nominal and did not affect the status and work of serving officers. A Colonial Research Service was formally created in 1949, mostly in the fields of agriculture, animal industry, forestry, biology, and fisheries. This was accompanied by the setting up of regional agricultural research stations, as in East Africa, West Africa, and the West Indies.

A large proportion of the staff were agricultural officers engaged in extension work. The headquarters of the Agriculture Department would in addition house specialists, such as entomologists, plant pathologists, and botanists or ecologists, who would tour the Districts on an advisory basis. It was common for a District agricultural officer to suppose that insects were eating his crops and call in the entomologist, only to be told that that it was a virus disease and required the services of the plant pathologist.

The earliest staff in soil science were appointed to posts called chemist, agricultural chemist or soil chemist, their primary task being to conduct soil analyses as a basis for fertilizer recommendations. A few of these chemists realized that if soil management advice, and agricultural extension in general, were to be done on other than a farm-by-farm basis, then soil maps were needed. It was through soil chemists turning their hand to survey that we owe the contributions of four of the pioneers: Frederick Martin in Sierra Leone, Geoffrey Milne in Tanzania, Arthur Hornby in Malawi, and A.W.R. Joachim in Sri Lanka.

Once the need for surveys had been recognized, appointments began to be made to posts which were effectively, if not always in title, Soil Surveyor. Among the earliest were Brian Anderson in Tanzania in 1947, and C.F. Charter, who was appointed Chief Soil Scientist to the Soil Survey Division of Ghana in 1949 and recruited his first three staff in 1951. Three soil surveyors took up posts in Nigeria in 1951, Colin Mitchell in Sudan the following year, and William Panton in Malaya in 1953.

A Colonial Pool of Soil Surveyors was created in 1954, probably at the instigation of Herbert Greene, Tropical Soils Adviser to the UK government. It was one of a number of such 'pools', intended to provide a source of expertise available to any part of the Colonial world. Four officers, already in service, were initially designated as members: Arthur Ballantyne, George Murdoch, Anthony Hodge, and Len Curtis. By 1960 the Pool had been expanded to ten, and shortly afterwards it was renamed the Colonial Pool of Soil Scientists.

The Pool supplemented, but did not replace, soil surveyors already serving with Colonial Departments of Agriculture. It was operated on the basis of medium-term secondments, typically two to three years, to specific countries. It helped to moderate the later system of 'experts' ('visiting firemen'), flown in to solve a problem and never seeing the country again. Rather, the Pool served as a basis for reallocation of officers once the projects on which they were engaged, typically reconnaissance soil surveys, had been completed. In 1964-65 the Colonial Pool was merged into the Land Resources Division.

The Land Resources Division

The Land Resources Division, to use the name it held at the height of its activity, rose from small beginnings to become the finest tropical land resource survey organization in the world. At a time when most colonies had acquired independence, it brought together a professional staff of eighty, with knowledge and field experience across a wide range of natural resource survey and management. Starting with inventory-type studies, surveys of soils, forests and pastures, it went on to develop an integrated approach linking resource survey, technology (agriculture, forestry, etc.) and socio-economic analysis. Between 1966 and 1987 the Division produced a series of thirty-six *Land Resource Studies*, which taken together constitute one of the greatest contributions to geographical knowledge produced by the United Kingdom. Whilst most activity took place in former British territories, the change in higher control from the Colonial Office to the Overseas Development Ministry meant that assistance was also given to non-Commonwealth countries, notably major studies of Nepal, Ethiopia and Indonesia.

Rarely can there have been an organization which so well illustrates the adage, 'tall oaks from little acorns grow'. Two such acorns were senior men in their fields: Brigadier Martin Hotine, head of the Directorate of Colonial Surveys, charged with topographic mapping of the colonies; and Professor Dudley Stamp, of the London School of Economics, the best-known geographer of his day. The third was a newly-qualified graduate in geography and geology, Martin Brunt.

In the 1930s, Stamp had organized the first *Land Utilization Survey of Britain*, to provide basic planning data for the government. After the war, he conceived the idea of setting up a World Land Use Survey, with the aim of producing global land use statistics from air photographs, starting with the British Commonwealth. Stamp approached Hotine asking for access to the Directorate's collection of air photographs, and for accommodation at the Directorate's offices at Tolworth, Surrey, for the first and only World Land Use Survey Secretary, John Callow. Meanwhile Brunt, having taken a postgraduate course in soil science under Walter Russell at Oxford, and declined a job with the Soil Survey of England and Wales, heard that the Directorate was offering a one-year post as Land Use Officer. Accepting this in 1956, he was assigned to a land use survey of the Gambia, comparing rice acreages on air photographs of 1946 and 1956. This was to be the first project of the nascent organization.

Brunt was joined by an Assistant Land Use Officer, Michael Bawden, at the time doing research on the geology of Spitzbergen. In 1959 they were amalgamated with the Forestry Air Survey Centre, to form the Forestry and Land Use Section of what had by this time been renamed the Directorate of Overseas Surveys (DOS). In 1964-1965 this was amalgamated with the Colonial Pool of Soil Scientists, to become the Land Resources Division of the DOS. The Division moved to accommodation in an unlikely headquarters for an organization with world outreach, Tolworth Tower, a multi-storey office building in Surrey. To complete the somewhat complex changes of name, the Division separated from the DOS, became the Land Resources Development Centre in 1978, before amalgamation with two other governmental research organizations as part of the Natural Resources Institute in 1987. Subsequently it suffered further cuts in staffing and budget, for some years operating on a reduced scale compared with the capacity during its heyday, its final project in land resources taking place in Namibia 2005-10.

The successive Directors were Philip Chambers, an ex-Colonial agriculturalist (1967-1972), Tom Rees, a forester (1972-1974), and Anthony Smyth, a soil scientist (1974-1987). Smyth had spent ten years as Soil Surveyor, Nigeria, followed by twelve as Land Classification Specialist to FAO, Rome. It was during his time that the major expansion in staffing took place. By 1984 professional staff numbered sixty-four, comprising a Scientific Section (soil science/geomorphology, ecology, hydrology, agriculture, forestry, land use planning and socioeconomics) and Specialist Support Units (cartography, remote sensing, data management, soils analysis, information and publications). The high degree of multi-disciplinarity enabled the Division to produce wide-ranging and innovative studies. It is a matter for regret that this outstanding organization, which could have remained a leader in British overseas aid activities, no longer receives the support to do so.

Consultant companies

In parallel with work by Colonial Departments of Agriculture and the Land Resources Division, consultant companies carried out surveys on a commercial basis. Funding came for the most part from international development agencies, principally the United Nations Development Programme and the World Bank. These were generally of the type called pre-investment surveys, where a development scheme is intended and surveys are required, firstly, to determine if it was viable, and secondly, to plan it efficiently. Frequently these studies were for irrigation schemes, involving high initial capital costs. Estimates of the economic returns, often based on crop yields and prices, and involving a somewhat contentious procedure of cash flow discounting, figure strongly in assessments of viability.

The first British company to undertake land and water resource studies was Hunting Technical Services. Its origins were on an exceedingly small scale. Immediately after the war, Pat Hunting had founded a company called

Hunting Aerosurveys, to provide air photography for map-making. In 1953, they realised the potential for mapping natural resources from air photographs, and sought to appoint an ecologist. The post was first offered to Duncan Poore, later to become a leading authority on tropical forests, but he turned it down and suggested his friend Vernon Robertson, at that time head of the farm belonging to the School of Agriculture at Cambridge University. Robertson was offered a two-year contract, and became a one-man Natural Resources Department of Hunting Aerosurveys, subsequently joined by a geologist, Greg Cochrane.

The years 1954-1957 were a make or break period for the new business. Robertson spent time travelling, trying to sell the services of the company. Their first major project was for irrigation development in Iraq; the engineering company, Binnie & Partners, called in Huntings to carry out soil survey, land classification and settlement planning. This was a critical project, in that it gave them a track record both in irrigation surveys and in the lucrative area of the Middle East.

Entry into Africa was made through the Elephant Marsh project, Malawi, and the Jebel Marra project, Sudan. Significant work in Asia included the Jengka Triangle land settlement project, Malaya, and the Lower Indus project, Pakistan. In the last-named, staff became caught up in the short India-Pakistan war. Much of the consultancy by Huntings was done in association with engineering companies. Organizational changes were completed in 1965 with the creation of Hunting Surveys and Consultants, with three operational divisions: Hunting Surveys, Hunting Geology and Geophysics, and Hunting Technical Services, the last headed by its founder, Vernon Robertson. For many years Harry Piper, a geologist, acted as coordinator of Huntings' natural resources staff.

The sheer quantity of work produced by Huntings in its heyday is vast. Their collection of natural resource surveys, 300 boxes full, was recently presented to the World Soil Survey Archive and Catalogue (WOSSAC) at Cranfield University. If the reports were stacked on bookshelves they would occupy over 100 metres of shelving. In addition there some are 20 metres of 15 x 20 cm standard cards for field site description.

The origins of Booker Agriculture International were very different. Booker began as a nineteenth century trading company in Guyana, owning sugar plantations and processing plants there. It expanded into a major agro-industrial business centred on food processing, mainly sugar but also other tropical crops. After nationalization of their Guyana plantations, they sought to establish new ones. The reason for setting up a Soils Department was that they had burnt their fingers with a sugar plantation in Nigeria on land which looked suitable but was anything but; the Managing Director is reported as saying, 'Never again will we do anything like that without a proper soil survey.'

Sugar cane is a demanding crop. It needs large amounts of water, yet adequate drainage. Being the highest producer, in terms of calories per hectare, of any tropical crop, it also needs fertile soil, together with appropriate fertilization. The Booker soils staff built up an expert knowledge of the land requirements for sugar, and when FAO set up a crop requirements database, the company was able to offer valuable guidance. Booker conducted land suitability investigations as part of general feasibility studies, some of which went on to become Booker-managed enterprises, for example Ramu Sugar in Papua New Guinea.

Arising out of their own need for standardized methods of survey and analysis, the soils staff collaborated to produce the *Booker Tropical Soil Manual* (1984), compiled by Rick Landon with major contributions from George Murdoch, whom colleagues referred to as a walking database. No comparable handbook had been produced by other institutions, and it became widely used outside of Booker.

Stimulated by competition and investment for development, agricultural consultancy firms multiplied. Among the better known based in Britain were W.S. Atkins, Landell Mills, and ULG. One such company (not among those mentioned) had a contract with a Middle Eastern state to establish and maintain trees in the desert, improving water-use efficiency by supplying each tree individually through drip irrigation. It was not foreseen that 'cups' of salt would build up around the roots, causing the trees to die. When they asked for more money to rectify this situation the response was that there was no provision for this in the contract. The company ceased operations and went bankrupt but, as is not uncommon in business, the staff set themselves up as a new company.

Irrigation engineering consultancy is an even more crowded field, a handbook for 1975 listing ninety-six companies. Among those which have carried out joint projects with agricultural consultants are Binnie & Partners, Sir Alexander Gibb, Sir William Halcrow, and Mott Macdonald. David Parry spent a career of forty years successively with Huntings, Lavalin International (Canada), and Mott Macdonald, starting as Soil Surveyor, subsequently a Project Manager, and ending as a Regional Director.

In commissioning soil surveys on a consultancy basis, there is a dilemma. The clients wish to ensure that they get value for money, and specify the number and spacing of traverse lines, soil pits and analyses. Grid survey, however, is an inefficient method, as compared with drawing of boundaries by air-photograph interpretation, and stratified sampling within the mapped units. Another problem is that soil microvariability, the range of properties over short distances, makes it hard to specify the accuracy of the resulting maps. A book by one of Huntings' leading staff, Stan Western, *Soil Survey Contracts and Quality Control* (1978) treats these and other problems of contract planning, survey execution, and reporting.

The reports of consultancy surveys are frequently bulky, running to many volumes, in part due to the practice of putting on record all the detailed observations. It is doubtful, however, if any can match a study by the Land Resources Division outside the Commonwealth, the Sumatra Transmigration project, the results from which were presented in eight regional suitcases.

Educational and national institutions

Many other institutions contributed to the post-war expansion of surveys. The Imperial College of Tropical Agriculture, Trinidad, continued a training role in tropical soils, later becoming the Agriculture Department of the University of the West Indies and transferring its research branch to the Regional Research Centre for the West Indies (RRC). Six British universities set up Departments of Soil Science or similar, among which Reading University developed special interests in tropical soils, and Edinburgh pioneered a Natural Resources Department.

Regional research institutes made important contributions, for example RRC, and the East African Agriculture and Forestry Research Organization

(EAAFRO) located at Muguga, near Nairobi. The first director of EAAFRO was the soil physicist, Bernard Keen, succeeded in 1955 by the then doyen of British soil science, Walter Russell. Crop research institutes, generally funded by a levy on exports, contributed to knowledge of soil requirements, mainly for perennial crops.

International institutions

Natural resources for agriculture are the responsibility of the Food and Agriculture Organization of the United Nations (FAO) based on Rome, through its Land and Water Development Division. It is not a funding agency nor does it have a mandate for research, yet besides conducting major field surveys under the aegis of other international bodies, FAO operated a series of compilation projects which were research in all but name. The most important in the early years were the *Soil Map of the World* (1970-80), which led to a degree of standardization in soil classification; the *Framework for Land Evaluation* (1973); and the *Agro-Ecological Zones Project* (1978-81), which led to an internationally acceptable method of agroclimatological classification.

When the Consultative Group on International Agricultural Research (CGIAR) was formed, the component institutes (in 2017 numbering fifteen) were mainly crop-based. Some carried out important soils research, notably fundamental work on soils for rice by the International Rice Research Institute (IRRI) in the Philippines. There was no international body responsible for tropical soils research until the foundation in 1985 of the International Board for Soil Research and Management (IBSRAM). The history of IBSRAM is a sorry one. Because the climate of opinion at the time favoured research into practical systems which farmers could immediately apply, it did not conduct centralized strategic research. Instead, it set up a set of regional research networks. Very briefly, it got the networking system wrong, failed to make substantial advances in knowledge, and was never accepted as a member of the CGIAR system. In 2001 IBSRAM was transferred to become a unit within the International Water Management Institute (IWMI), Sri Lanka, since which time its soils research has been greatly reduced.

Despite laying much emphasis on sustainability, the World Bank (Washington DC) was never a force in land resource development, other than through its Forestry Department. Its very few natural resource-based staff, among them a soil scientist with long experience in Malaya and the West Indies, John Coulter, struggled against the dominant ethic of appraising development in terms of economic returns. Although peripheral to the present story, I cannot resist from quoting from an internal World Bank review of project performance (which may have been confidential at the time). This contained a formula for an *unsuccessful* project, which would contain any or all of the following:

- An overall regional development plan intended to reduce poverty, without a sectoral focus.

- A project to improve nomadic pastoralism.

- A project which depends to a significant extent on cooperation between two or more government agencies.

Methods of survey

Air-photograph interpretation

To look at a landscape on air photographs is a revelation. Only the most simple and inexpensive equipment is needed, a mirror stereoscope to allow a pair of photographs to be viewed in three dimensions. The photographs need only be in black and white, most usefully at scales between 1:25 000 and 1:40 000. When this is done, the landscape is displayed before your eyes as never before: the landforms in somewhat exaggerated relief, together with the patterns of vegetation, land use, and settlement. It shows far more detail than can be found on any map, and a better overall appreciation even than travelling across the ground. In particular, air-photograph interpretation allows the surveyor to distinguish recurrent patterns, and draw boundaries between them. This was the greatest advance in survey methods in the period following the Second World War.

The first use of air photographs for resource surveys is sometimes ascribed to the United States in the early 1930s, but there had been earlier trials. These were the work of two individuals whom we have already met in senior positions but like all such, were once keen and enterprising young men. Dudley Stamp was cutting his geographical teeth in Burma. In 1924 he and a group of forestry officers mapped the forests of 3500 km^2 of the

Irrawaddy Delta. Stamp was subsequently to write one of the most graphic book dedications which field staff make to their wives, 'In memory of bullock cart days and Irrawaddy nights.'

In the late 1920s, Martin Hotine recognized the potential of air photographs for topographic map-making, as a far more rapid means of mapping detail than the traditional method of plane tabling. In 1928 an experimental run of photographs was taken across Lusaka, Zambia, and in 1935 the capital was moved there from Livingstonia. The 1928 photography was interpreted by Colin Trapnell and R. Bourne who clearly perceived the potential of 'aerial survey in relation to stock taking of the agricultural and forestry resources of the British Empire.' Considerable experience of air photography was acquired from reconnaissance flights during the Second World War. The Allied Central Interpretation Unit (ACIU) at Medmenham was the headquarters of Photographic Intelligence, and most operations, by air and by land, made use of target material prepared at the Unit. Enemy troop movements were monitored. Not without difficulty, the research centres and launching sites of the V1 ('doodlebug') and V2 rockets were identified. Before the D-Day Normandy invasion, hourly flights over the Normandy landing beaches allowed the construction of beach profiles from successive positions of the strand line (for the purpose of deception, an equal number of sorties were made over Calais). Some years later the government was about to pulp the large collection of wartime photographs when Professor Stanley Beaver of Keele University heard about this. After prolonged discussions over security aspects, Keele acquired the collection of five million photographs, a kind of Domesday record of the European landscape during the 1940s. This, and much other material, has since been transferred to the National Collection of Aerial Photography, Edinburgh. It was this wartime experience of stereoscopic air-photograph interpretation which led to its extensive post-war use for mapping and land resource survey, starting with flights by the RAF in Africa in the late 1940s. It was also the germ for the foundation of Hunting Aerosurveys.

Multi-spectral scanning from satellite imagery began with the launch of the first of the Landsat series in 1972. It was hailed as a breakthrough, and numerous articles on its potential appeared, far more than cases of its use in practice. It has opened up great opportunities in the mapping of land use and vegetation, especially in monitoring changes over time, for example forest clearance. It is valuable for providing an overview of landscape patterns over large areas, more efficiently than by air-photograph lay-downs. Techniques are being developed to monitor soil degradation and even soil type always, of course, in conjunction with field observations. Unfortunately there has also been a negative effect, where in planning projects, high-level officials say, 'We don't need a soil survey, we will use satellites.' This is false, and regrettable. Whenever it is intended to change land use or management, there is no substitute for going into the field and finding out what is there.

The land systems approach

Three organizations with different objectives, working independently, hit upon the same technique. The first was the Australian government, which was concerned about the use (or rather lack of use) of the sparsely occupied lands in the north and interior of the country. The government body for research, the Commonwealth Scientific and Industrial Research Organization (CSIRO), set up a Division of Land Research and Regional Survey, with the aim of making an inventory of the land resources of these areas. The first of their Land Research Series was *Survey of the Katherine-Darwin region 1946* (published 1953). The second organization was the National Institute of Road Research, South Africa, seeking information on soil-engineering properties for the design of road systems. The third was a unit of the British army, the Military Experimental Engineering Establishment (MEXE) who wanted to be able to forecast whether tanks would sink into soft ground, and other military questions, without on-site access. On learning of this common experience, these organizations came together in 1966 and a standard terminology was agreed.

The reason for this convergence of methods is apparent to anyone who tries air-photograph interpretation. Natural landscapes are quite clearly differentiated at two scales. First there are recurrent patterns, covering tens or hundred of square kilometres, usually extending across many individual photographs. These are patterns of the total landscape, landforms, vegetation and, often, land use, generally determined by the underlying geology coupled with the geomorphology, such as the degree of dissection. Early geographers had some premonition of these regularities as perceived from the ground, calling them natural regions. The term employed by CSIRO and subsequently adopted worldwide was the *land system*, a recurring pattern of landforms, soils and vegetation, and having a relatively uniform climate.

Land systems are made up of *land facets*, areas within which for most practical purposes environmental conditions are uniform. They are much smaller, of the order of hundreds of square metres. Examples are the top of a plateau, a steep escarpment, a valley side or a swampy valley floor. The same land facets are found across the whole of a land system; indeed, an alternative definition of the latter is a recurring pattern of land facets. Spatial units at intermediate scales have been distinguished.

As the news spread, the land systems approach was very widely adopted. In reconnaissance surveys, only the land systems are drawn upon the map. Each system is usually shown as a block diagram, accompanied by a table which shows the geology, landform, hydrology, soils and vegetation of each of the land facets. The actual or potential land use is often added. Cases where land systems have been broken down by mapping their individual land facets are rare.

There is a problem: soils are hidden under the ground, and cannot be seen on air photographs.[1] This is the same difficulty that faces a soil surveyor in the field. Consequently soil survey is heavily dependent on finding relationships between landforms, vegetation, and the underlying soils. Following initial air-photograph interpretation, the main function of field survey is to establish these relationships by means of soil profile descriptions on each of the land facets.

Tropical soil science

Up to 1940, most knowledge of soils had been built upon experience in temperate latitudes. Notions of tropical soils could not be other than simplified, sometimes conjectural or erroneous, since so few scientists had seen them.

The years following the Second World War saw a massive advance in knowledge, the driving force of which was soil survey. Surveyors found that the complexity of soil properties was far greater than had previously been

---

[1]   Recent developments in multi-spectral scanning have made some progress in remote identification of soils, including land degradation.

supposed. They sorted them into soil types, and at the same time discovered their relationships with the soil-forming factors: climate, drainage, geology (parent material), vegetation, and the impact of land use. This led to an insight into the soil-forming processes, the key to understanding how observed distributions had come about. Conversely, once the soil-forming processes and factors present in a given area had been worked out, the survey itself became more efficient. Discussion about the evolution of tropical soils, at conferences and in publications, made the subject more exciting than it is today.

An early task was soil classification, grouping soils into a limited number of discrete classes not only for the purpose of mapping but as a basis for management decisions, in modern jargon, technology transfer. Everyone could agree on the existence of certain soil types. The heavy black clays which swelled up when wet came to be called vertisols. Many regions had large areas of thin and stony soils, and it mattered not whether one called them that or dignified them with the name lithosols.

The problem came with what can be descriptively called the red and yellow soils of the humid and sub-humid tropics. Some were moderately fertile, a few could be cultivated almost continuously, others required long periods of fallow. For a time, countries drew up their own national classifications. The Ghana system devised by C.F. Charter, of ochrosols (more fertile) and oxysols (less fertile) had much to commend it, but was not employed outside West Africa. More widely adopted was the system devised by the Commission for Technical Cooperation in Africa (CCTA), led by the Belgian J.L. d'Hoore as the basis for the first *Soil Map of Africa*, in which the red soils could be ferruginous soils (moderately fertile), ferrallitic soils (infertile) or ferrisols (the highly fertile soils formed from basic rocks). Discussion ranged between natural classifications, centred upon recognizing 'ideal' soil types in the field, and artificial classifications, based on precise defining values of chemical properties. As a review article on soil classification put it in 1967, 'Scientists who are otherwise reasonable and unemotional are liable to behave quite differently when discussing this topic.'

In 1961 FAO embarked on a twenty-year project to draw up a *Soil Map of the World*, for which they needed a standard set of names into which national classifications would be converted. The red and yellow soils were placed into ferric luvisols, acrisols and ferralsols. FAO insisted that this was a legend to a soil map, but many adopted it as a classification system. Concurrently, the United States Soil Survey, the longest-established and largest in the world, produced a classification which for fifteen years was known by the strange name of the 'seventh approximation' but eventually appeared as *Soil Taxonomy*. It is highly complex and has a forbidding nomenclature; all previous soil names were rejected, so that the definitions of the new classes were those in the taxonomy itself. For a substantial period it was uncertain whether the FAO or US systems would be most widely adopted, and some national surveys sensibly carried conversion tables of local soil classes into each of them. FAO then diplomatically worked away at reconciliation between the two systems, adapting some of their defining criteria to make them compatible. The outcome became the *World Reference Base for Soil Resources*, recognizably evolving from the earlier FAO classification. The US taxonomy however, is still in concurrent use.

The higher-level classifications are for soil scientists, in order to know that a soil found in one country is the same as one which they know from elsewhere. What matters for practical purposes is that there should be classes and names which are recognizable in each country, not only by soil scientists but also by agricultural extension staff and farmers. These typically take the form of a local name, generally where the soil was first identified, followed by a summary description in plain language, e.g., 'Lilongwe series: a deep, reddish brown to red, sandy clay or clay, well-structured and friable, giving good rooting properties, and moderately fertile when well-managed.' This conveys the information which local agriculturalists want. Only the research staff need to be told that it is a ferruginous soil (CCTA), ferric luvisol (FAO), or alfisol — ustalf (US taxonomy).

Carrying out a soil survey

'Give us the tools and we will finish the job.' The phrase belongs to Winston Churchill in a radio broadcast of 1941, asking President Roosevelt for United States aid to the British war effort. It suitably describes the situation in which Colonial soil surveyors found themselves in the early post-war period. The desire for land development and the institutional basis had been set up. Air-photograph interpretation was available early on, although some early post-war surveys were still done by traverses. Other tools, such as knowledge of tropical soil formation, were not so much 'given' as developed interactively between field surveyors and academic research, including by movement of staff between the two. Methods for soil profile description were evolved primarily in the United States, adapted for tropical conditions by FAO handbooks. A summary of the state of the art was given in a book that David Dent and I wrote jointly, *Soil Survey and Land Evaluation* (1981).

So how does one carry out a soil survey? This very much depends on the scale and objectives, but for a reconnaissance study the usual steps followed were as follows:

- Collect topographic base maps, of whatever quality available.

- Commission, if necessary, the taking of air photographs. Assemble these and (occasionally with some difficulty) find out where they lie on the maps.

- Using a stereoscope and a glazed-surface crayon, draw boundaries on the photographs around all repeating patterns of landscapes, following them from one photograph to another and reconciling problems.

- Now set out in the field with a survey team, photographs in hand. It is usually more efficient to make use of all roads and tracks, supplementing this by compass traverses across the bush (or through the forest) where necessary. In forested land, traverse lines must be cut.

- Take field observations of two kinds, traverse and site. When travelling, make a continuous record of landforms, vegetation, soils and land use, and note the points at which these change. Stop, describe the environmental conditions and the soil profile, using a pit, auger boring, or road cutting. Through your interpreter, talk to as many farmers as possible, asking their views on the potential and problems of their land.

- Return to the office and collate all this information. Decide which distinct soil types can be recognized, and how these are related to the boundaries previously drawn on the photographs.

- Proceed to a stage of interpretation in terms of land potential. Bring in the knowledge of agricultural extension officers, and relate soil or land system map to land use potential.

- Present the results: draw up the map or maps, and write the covering memoir, including the consequences for practical land management and development.

This is a simplified account of what was actually done in many surveys. Two further points may be added on what ought to be done. Firstly, during the field survey, an effort should be made to visit sites where agricultural performance data are available: experiment stations, crop variety or fertilizer trials, plantations which have kept yield records, and farms for which there are records by agricultural marketing boards. Profile pits should be dug at these sites, so that the performance data can be related to soil types.

Secondly, when the initial survey is finished, at least one of the surveyors should not go away. They should continue to make recurrent soil observations in the region, observe the success or problems of changes in land use, and lend their soils knowledge to project management and extension staff. In the longer term, they should set up a system of soil monitoring to detect fertility decline and thereby sustainable land use. Regrettably, this has rarely been done. The accident of political circumstances is partly to blame. Reconnaissance surveys were mostly done during Colonial rule, and the staff who made them departed soon after independence. On major development projects, however, there is no such excuse. Now that the staff of national soil surveys are mostly local, they are in a better position to carry out monitoring and advisory work. Departments of Agriculture should employ soil scientists in the same way as entomologists, plant pathologists and other specialist staff, as ongoing technical backup to field advisory staff.

It would be wrong to leave soil survey without mentioning a most vital piece of equipment, the Land Rover. It is safe to say that a high proportion of surveys during this period, when not conducted on foot, were made in this vehicle. The Land Rovers of the 1950s and 60s were very different to the comfortable Range Rover of today. The seats and the springing were hard, a metal ledge would cut into the passenger's knees, and there was no synchromesh in first and, in the early models, second gear. But they were unmatched when it came to traversing potholed roads, tracks, and driving straight across the bush. They were also easy to maintain, with an accessible engine; my driver once replaced a broken fan belt with bark from a (carefully selected) tree. The original Land Rover, registration number HUE 166 or 'Old Hughie', can be seen in the Land Rover Heritage Museum in Gaydon, Warwickshire, UK. The sight of it will bring to Colonial officers of that period fond memories, and perhaps a pang of pain in the knees and back.

Interpreting surveys: land evaluation and land use planning

How does one take an area mapped as having a ferruginous soil, or the Lilongwe series, and put across to users of the survey the fact that this is an excellent place to grow maize and groundnuts, but will need nitrogen fertilization? One way is to say exactly that. Part of a survey report can be in practical, descriptive terms, which can be understood by extension staff and the farmers whom they advise. This can be a good way to present the interpretative part of a report, and some good examples are described in the regional accounts below.

However, the unhelpful, jargon-filled presentation of other survey reports led in the 1960s and 1970s to calls for some standardized system of interpretation. The first of these was land capability classification, not so much an interpretation of soil surveys as a field survey method in its own right. It originated with the US Soil Conservation Service, as a basis for farm planning. Capability classification is based on the concept of limitations, land characteristics which adversely affect land use. Land is assigned to eight capability classes, ranging from Class I, land with few limitations, to Class VIII, land with limitations that restrict its use to wildlife, water supply, recreation, or aesthetic purposes. In between come classes assigned to grazing and forestry. Adaptations of this system were devised in a number of countries, notably Zambia, Zimbabwe and Malawi, and successfully applied to set out new farms on previously unsettled land. Two problems which arose were the inbuilt hierarchy of uses, meaning that forestry was assigned only the worst land, and slope limits for cultivation which in areas of high population pressure on land are often unrealistic.

The second system was the US Bureau of Reclamation's system of Land Classification for Irrigation. This assigns land into six classes according to suitability for irrigated agriculture. Starting with physical land limitations of slope, soil and drainage, it is supposed to convert these into economic terms, leading to land classed according to its forecast ability to repay the capital and recurrent costs of irrigation. Although irrigation schemes as a whole are always appraised in economic terms, the Bureau of Reclamation system has been less widely applied outside the United States.

Conscious of problems in using land capability classification, FAO sought to devise a standardized method of appraisal, which was called Land Suitability Evaluation, or simply Land Evaluation. The origins of the approach they were to adopt lay in work by two Dutch scientists, Klass-Jan Beek and Jacob Bennema. Their pioneering work was discussed and modified at an 'expert consultation' at FAO in 1973, following which two consultants, Robert Brinkman (Netherlands) and myself, were commissioned to compile an outline. This became the *Framework for Land Evaluation* (1976).

The key principle of land evaluation is to define the purposes for which land is being assessed. These could have been called the land use types, although to make it clear that these were not the actual uses of land but uses being assessed, the term 'land utilization types' was adopted. These might be individual crops, or use for forestry or grazing, or a land utilization type defined in some detail, e.g. oil palm cultivation by smallholders, with low inputs (which has different requirements from cultivation on plantations with high fertilization). There are four land suitability classes: highly, moderately, and marginally suitable, and not suitable. Land is assigned to these classes separately for each defined land utilization type. The outcome is a map which answers two questions:

- We have a type of use in mind: where are the best areas to practise it?
- We have an area of land in mind: what types of use are best for it?

It is noteworthy that long before the modern attention to sustainability, one of the six 'principles' which guide the *Framework* was given as, 'Suitability refers to use on a sustained basis.'

The *Framework* became one of the most widely used of all FAO publications. A note may be put on record about its origins. Robert Brinkman and I agreed that I should write an introductory chapter, outlining the method, and we would then meet to discuss who should draft the rest. When

Brinkman saw the 'Introduction', he remarked, 'Wait a minute. We don't need very much more. With a little bit of expansion, and the addition of examples which I will find, this will be adequate to put across the method.' That is the reason it is such a compact document. Detailed methodology followed in later publications on land evaluation for rain-fed agriculture, irrigated agriculture, extensive grazing (the most difficult to assess), and forestry.

By converting technical soil and environmental descriptions into maps of land suitable for maize, cocoa, irrigation, forest plantations, etc., it marked a large step forward in making results accessible to users. The fact that it began by defining land utilization types, new uses for the land which could reasonably be considered, was also an advance. There have been cases, however, where the method seems to have been applied as an end in itself, mapping suitability classes without regard for the practicalities of land development.

Having achieved a method for land evaluation it was logical for FAO to try to formulate a similar set of procedures for land use planning. This was not wholly successful. A committee in the Netherlands had tried to do it and came up with a list of 'basic steps in land use planning': 369 of them! The FAO *Guidelines for Land-use Planning* (1993) reduced this to ten basic steps, and is probably as good a systematisation as is possible. By the time this appeared, however, it was being questioned whether planning, in the sense of top-down direction, should be attempted at all. The participatory approach, with emphasis on farmers, institutions and other stakeholders, had gained favour, in part because improved use of land already occupied was by then so much more common than new land settlement. A new FAO publication, *Negotiating a Sustainable Future for Land* (1997) sets out this approach, subsuming the Framework and the Guidelines in seven pages. The problem with setting out a formula for land use planning is that the objectives are so varied. The basic approach has to be, find out the problems of the area, realistic ways of solving them, talk to people and institutions, then collect the data required to achieve this. This can be very much more effectively achieved when it is built upon a foundation of previous systematic resource surveys.

4

EAST AFRICA

FINANCED BY THE Colonial Office, staffed by newly-appointed soil surveyors and ecologists, and with the vital tool of air-photograph interpretation to hand, the ground had been prepared for the rapid expansion of land resource surveys. During this time, reconnaissance scale surveys were undertaken in most of the territories, with complete national coverage achieved in some. A few soil chemists remained in post during the Second World War, although many were seconded to military duties. For the most part, however, survey activity followed a period, surprisingly short, of post-war recovery and re-staffing.

Botanists or ecologists sometimes preceded soil surveyors in Departments of Agriculture. Two early post-war appointments as ecologist were George Jackson, Malawi, 1950, and Ian Langdale-Brown, Uganda, 1953. Both were graduates from Colin Trapnell's Ecological Training Scheme based on Oxford and Kenya.

One of the earliest post-war appointments in soils was John Coulter, Soil Scientist, Malaya, in 1948, mainly concerned with fertility studies but also having survey responsibilities. Most early *de facto* surveyors were in posts as soil scientist or chemist. The earliest to be called soil surveyors were probably the founding members of the Colonial Pool in 1954, followed by Cliff Ollier's appointment to Uganda in 1956.

The early to mid-1950s saw the start of reconnaissance survey activity in several countries, and by 1970 a substantial number had been completed and published. Surveys at the small scales of reconnaissance work, and covering large parts of a country, then became less common. The first reason is that when a survey has been completed there is no need to do it again. A second was a swing of opinion away from reconnaissance studies, and towards more detailed surveys for specific development purposes such as irrigation or land settlement schemes. These roughly coincided with the period when most colonies acquired independence, 1956-1966 for the larger countries, extending past 1980 for the smaller, mostly island, territories. Independence was often preceded by a period of civil disturbance, whilst the newly-empowered governments had other immediate priorities. So we may take the period from 1950 to 1980, extending later in some countries, as marking the golden age of reconnaissance surveys.

With the explosion of staff numbers, the biographical approach employed for the pioneering period is no longer possible, although a few individuals with long and distinguished careers are given a separate chapter. For the most part, the treatment will be by country, taken regionally.

Uganda

With the first designated soil surveyor, and the earliest post-war completed national survey, there is no better country to begin with than Uganda. The national reconnaissance surveys were the most successful of any Colonial territory, both for completeness and achievement within a short period. The decision to carry out the surveys was taken, staff appointed, the facilities and budget made available, and results compiled and printed, all within the space of six years, 1956-1962. Credit for this must go in the first instance to Ernest Chenery, appointed Chemist in 1950 and becoming Chief Research Officer in 1958.

Chenery's degree was in agricultural chemistry but he had spent time in the field with Gilbert Robinson, founder of the Soil Survey of England and Wales in the 1930s. He foresaw the need for a survey of the land resources of the country, secured the funding, appointed staff, took technical advice from them, and on completion wrote a summary account of the soils himself. His original concept was ambitious: surveys of the soils, vegetation, and farming systems, each consisting of six volumes, one for each province. There were also to be accounts of each of Uganda's main crops, as well as pastures and animal husbandry, but only the memoir on sorghum was completed.

The proposal to carry out a reconnaissance soil survey was taken by the Ugandan Agricultural Production Committee in 1954. The intention was to gain a better understanding of the natural resources on which to base a strategy to improve agricultural output, both of local food crops and the valuable perennial crops, especially coffee and cotton, exports of which were already bringing in £400 million.

Three soil surveyors took up posts in 1956: Cliff Ollier, John Harrop and Stanislav Radwanski. They divided up the country by mutual agreement. Harrop was a mountaineer so was given the Western Province, allowing him to put his hobby to good use by ascents of the Ruwenzori. Radwanski was the only married man among them so was given Buganda, the area around the capital city. After these were allocated, Cliff Ollier took the northern and eastern plains, 'a huge area, which did a lot for my mileage allowance.'

This left Karamoja District, occupied by people whom the harsh environment had rendered warlike. It was a restricted district, for which permission was needed to enter, and good relations with the locals of some importance. The problem was solved by enlisting the help of John Wilson, the Agricultural Officer resident in the District, speaking the language and with the confidence of the people. He was taught the rudiments of both vegetation and soil survey becoming, with his training in agriculture and linguistic proficiency, a veritable polymath, comparable with Uganda's Special Commissioner in 1899, Sir Harry Johnston.

The unofficial leader of soil survey activities was Cliff Ollier, who had a high capacity for fieldwork. He reached the country, travelling by ship to Mombasa and up-country by rail, in January 1956. Ollier started with a geology degree from Bristol, took a Colonial Office Post-Graduate Studentship in Soils as an alternative to National Service, and was sent to Rothamsted, headquarters of the Soil Survey of England and Wales, for a year's training. He worked mainly with surveyors in the Chilterns, a chalk landscape hardly comparable to African conditions but which at least had hills and valleys allowing survey to be based on geomorphology. Only very late on did he discover that he was supposed to be supervised by Herbert Greene, by which time it was too late for Greene to pass on his considerable knowledge of tropical soils, so Ollier reached Uganda free of preconceptions.

Cliff Ollier explains the purposes of soil survey to the
Karamajong people, Uganda

'As I was the first arrival in Uganda, I started to make the rules', he writes.
He worked out the area to be covered, the length of roads, and calculated
(to Chenery's surprise) that they could afford to make an observation only
every four miles (6.4 km), with special traverses in significant areas. Some of
the base maps were too poor to use, and they requested rapid production of
air-photograph print lay-down maps. Arriving in September/October the
same year, 'John and Stan seemed happy to carry on the same way'.

The Uganda civil service had a large complement of botanists, four in 1954
and eight by 1966, but these were mainly concerned with coffee and cotton.
The only full-time botanist on ecological survey was Ian Langdale-Brown,
arriving in 1953. Langdale-Brown completed the ecological survey more or
less single-handed, although 'my wife acted as my unpaid field assistant', a
not uncommon practice in the service. The ever-adaptable John Wilson was
again recruited to cover Karamoja District.

The maps of vegetation were produced independently of the soil survey.
Langdale-Brown did not work jointly with the soil surveyors in the field, nor
did they deliberately employ boundaries in common. This decision was

somewhat surprising in the light of the integrated surveys pioneered by Trapnell, but was considered by Chenery and the respective surveyors to lead to independent results. The Uganda work was completed by a survey of indigenous farming systems by Dennis Parsons.

The results of the Uganda surveys were published as *Memoirs of the Research Division*: Series 1 on soils, Series 2 on vegetation, and Series 3 (incomplete) on indigenous agricultural systems. These are workmanlike bound volumes, systematically arranged and clearly written, and inexpensively produced. The maps are at 1:500 000, a smaller scale than that of many reconnaissance surveys but one which allowed the work to be rapidly completed. This scale also allows maps of a size convenient to use, on a desk or in the field. Ollier suggested in retrospect that it might have improved efficiency if more initial use had been made of terrain classification, the landform basis of land systems.

The soil maps are in black and white only and with sparse base-map detail, making it hard to obtain an overview of the distribution. On completion of the survey, however, they were combined into a colour-shaded map of the whole country at 1:1 500 000 scale (1967). For vegetation, as Lang-dale-Brown had himself completed five of the six provinces, he was more readily able to consolidate them in *The Vegetation of Uganda and its Bearing on Land-use* (1964). This includes 1:1 500 000 scale maps of vegetation, current land use, range resources, ecological zones, and rainfall. The Uganda surveys are the most comprehensive and complete of any of the larger British territories.

In the same tour of service Ollier did a number of local surveys, 'when the admin realised I could produce a readable report on time.' One job was what would now be called an environmental impact study, of the effects of the proposed Mutir Dam (an alternative to the Aswan Dam), using the boat that had appeared in a film role the previous year in *The African Queen*. One evening his colleague on this study, John Sutcliffe, decided to have a formal dinner and invite the Captain of the Lake Albert Steamer:

> We were in a thatched rest-house, it was incredibly hot, the mossies were the most ferocious I've known, and bats were s----ing from the roof. Eventually a messenger arrived through the pouring rain bearing a message in a cleft stick (the only time I ever saw one)

saying the Captain apologized that because of the weather he was unable to accept our invitation. So we had dinner alone, in evening dress, black tie and mosquito boots!

Like many tropical soil surveyors, he occasionally encountered hostility from Africans who supposed he was making surveys so whites could come and take their land. Others thought he was putting something *into* the soil. Why else would he take the trouble to fill in the pits — with poison so they would die, a contraceptive potion so they would have no children, or jigger parasites that would make them lame so that they could no longer resist British soldiers (there was an historical precedent for this unlikely supposition)? Many of us have occasionally run into this problem but it is rare, can nearly always be diplomatically explained, and is wholly outweighed by the friendly welcome that is the norm.

In the course of survey work Ollier developed ideas on tropical weathering which were to set him up for an academic career. Visiting him in Perth, I enquired about his hobbies, what he did when not working. The reply was, 'Well, in one sense you could say I don't have any hobbies. But in another, I'm lucky enough to have spent my whole career doing what I enjoy.'

John Harrop stayed on in Uganda after independence in 1962, turning his attention to the use and management of soils. He tried, but narrowly failed, to obtain a Ugandan successor. It was his regular practice to take final-year students from Makerere University on annual field trips to show them the range of soils in the country:

> It was my hope that these outings might have identified a candidate for the post of soil surveyor. One excellent student did appear, but unfortunately failed to get an upper second (necessary for entry to the Research Division), only because he fluffed his economics paper. Despite pleas from Harrop and Professor Harry Birch the examination board would not allow compensation for the qualities he demonstrated in the soils papers, so we lost him.

Harrop stayed on to become Chief Research Officer. In 1971 came the coup in which General Idi Amin Dada seized power:

> Our new Minister, Fabian Okwari, told me that the President had specifically requested I should stay in Uganda, but the British High

Commission opposed this and I had to decline the request as tact-
fully as I could and hope there would be no repercussions.

He left Uganda in 1971 and joined FAO, working on a land capability
project based on the Soil Research Institute at Bogor, Indonesia. 'Being in
Uganda was the best event of my life' writes Harrop. He met his wife,
married and had two children there, and enjoyed 'a great social life with lots
of sport.' He brought his golf handicap down to seven, surely a record for
field survey staff who have less time for social and sporting activities than
most officers.

Ian Langdale-Brown, after his time in Uganda, spent four years conducting
ecological surveys for the Land Resources Division, before going on to a
distinguished research career at the Department of Forestry and Natural
Resources of Edinburgh University.

We cannot leave Uganda without drawing attention to a remarkable study
of 1938 by the Department of Agriculture, coordinated by its Director of
Agriculture, John Tothill, with the lengthy but informative title, *A report on
nineteen surveys done in small agricultural areas in Uganda with a view to ascertaining
the position with regard to soil deterioration*. This is a veritable Domesday Book
with a breathtaking amount of detail. For each 'hundred' (village administrative unit) studied, he gives the population and its density, soil types and
their condition, pastures and condition, areas under each crop, cattle and
goats, occupations, diet, and much else besides. This is followed by lists of
individual farmers by name, their families, occupations, religion, clan, sub-
clan and 'totem', the clans of their wives and parents, how many blood
brothers and whether they consult a witch doctor. Then there is a wealth of
detailed economic information, e.g. that the village of Bunjenje in Bunyoro
District receives 4314 shillings from the sale of tobacco, and a likewise
specified income from many crops and livestock products, down to 4 shillings from the sale of eggs. Twenty-three men are given exemption from
paying tax because they are retired *askaris* (soldiers), or old, sick, lame, or
exempted for one year because wives died.

There is a fabulous research opportunity for a follow-up study, going back
to these same villages, looking into the present population density, agriculture, and condition of land resources, finding descendants of the same individuals and comparing their farm sizes in 1935 and today. A gold mine for

anthropologists, it should not be left to them alone, but needs historical geographers or more precisely, historical agriculturalists and land resource scientists.

Kenya

Kenya differed from most colonies in having a substantial settler-farming element. Relations between the two were sometimes strained. Government focused on the welfare of the Africans, believing that the settlers could look after their own interests. The settlers, led by the diminutive but energetic Lord Delamere, felt they were being neglected. It was all very good-natured. The attitude was that development was a private matter, best carried out by individual farmers on the basis of trial and error, supported by analyses carried out at the Scott Agricultural Laboratories. For advice on soil management one could call upon G.H. Gethin-Jones, Soil Chemist 1931-49, a length of service and depth of experience not often matched in later times.

There is an agricultural leaflet from 1913 by V.H. Kirkham, *Hints on collecting soil samples*, which probably refers to sampling for determination of fertilizer requirements. *A preliminary survey of some of the soils in Kenya* by D.S. Gracie, published in 1930, is a substantial report giving descriptions and analytical data District by District, mainly those of the 'White Highlands' around Mount Kenya and Nairobi. This was a valuable guide to soil management for field officers and settler farmers but, lacking a map, not a survey in the normal sense. Concern with erosion was shown from the 1930s onwards, and a Soil Conservation Service was in place by then. Research staff in 1932 comprised a mycologist, a plant breeder, two soil chemists, and no less the five entomologists, a preoccupation with what was eating your crops that was commonly found in early Departments of Agriculture.

In the early post-war period, for national soils coverage recourse had to be made to Milne's 1936 map of East Africa. Then in 1959, the accumulated knowledge of a range of government officers was put into the *National Atlas of Kenya* (1959), a compilation of forty maps at 1:3 000 000 scale. These include climate, soils, water resources, vegetation, ecological potential, agriculture, forest reserves, national parks and a range of social information. The soils map, by J.E.D. Robinson and R.M. Scott, is based on descriptive units of the kind, 'Very dark brown clay loams', some mapped as striped soil associations; it is not a new survey but a compilation of existing knowledge. The maps of vegetation and ecological potential are by the leading authority on East African pasture resources, D.J. Pratt. There is a wonderful collection of early maps from 1596 onwards, and a 'coloured spaghetti' plot of the routes of nineteenth century explorers. National atlases were in fashion in the 1960s, although nowadays they sit gathering dust on library shelves. It is hard to assess how useful they were, though some were deemed worthy of several editions. Certainly they were valuable for educational purposes, and perhaps as a readily accessible source of background information. They may have served a purpose in drawing attention to what was not known, for it is only the compilers who appreciate just how misleadingly precise are the beautifully printed coloured maps.

In the post-war period, the Soil Survey Unit directed its attention to local surveys for specific development projects. Fred Collier was the first Soil Surveyor 1958-61, leaving to become Head of Soils at the consultant company, Huntings. John Makin was Head of the Soil Survey Unit 1963-68, and in 1967 was joined by the first of a distinguished line of Kenyan soil surveyors, Nelson Nyandat. Surveys at detailed scales with a focus on suitability for bananas, tobacco and coconuts appeared in the 1960s, many carried out by Makin. In 1974-76 a series of studies was made of irrigation suitability along the River Tana. Kenya is unfortunate in having three-quarters of its surface dry, but lacks a major river from a mountain area — the Tana is not on the scale of the Nile or Indus. Notwithstanding this, a line of small irrigation schemes was put in place, although the subsequent management of these left something to be desired.

Before continuing with the main story, two activities which were not followed up can be noted. The first originated from the UK Military Engineering Experimental Establishment (MEXE), located in Hampshire. They wanted to assess the suitability of terrain for tanks, foundations of roads and airfields, sources of materials and the like, in a military situation in which there was not access to the terrain. Richard Webster joined the Oxford-based team investigating this in 1961, and, 'by 1964 we had satisfied ourselves and the Ministry of Defence that a physiographic classification of land was in principle a sound basis for these purposes. The Ministry wanted to test the idea in a potentially active theatre' (for military operations —

Kenya was for long used as a training base for the army). They invited geomorphologists and soil scientists from Swaziland, Tanzania, Zambia and Kenya, with the aim of testing the scheme on a large scale. 'However, the politics of the post-Colonial era were changing rapidly' writes Webster:

> It was soon evident to the military that it could not do tank trials and the like without embarrassing the host countries. It was also clear that the main benefits would be peaceful: building ordinary roads, planning land development for agriculture, so the emphasis shifted. The money came out of the military budget, but with the knowledge and support of the Ministry of Overseas Development.

The outcome was *A Land System Atlas of Western Kenya* (1971). With a map at 1:500 000 and over 300 pages of land system descriptions it was in many ways a great advance in reconnaissance survey, yet does not seem to have passed into the mainstream of knowledge.

The second supporting activity was mapping vegetation, although in Kenya this turned out to be something of a blind alley. Since the fieldwork in Colin Trapnell's ecological survey training was carried out in Kenya, it seemed natural to extend this into a survey. Trapnell, by then in his late seventies, was joined in this by Martin Brunt of the Land Resources Division. In 1987, eight map sheets showing vegetation and climate at 1:250 000 were produced, but their time had passed. Ecological survey based on vegetation, highly appropriate in pre-war Zambia, was not relevant to the farming conditions, agricultural and pastoral, of modern Kenya. As regards vegetation, the country's leading post-independence survey need was to monitor changes in forest area, as a basis for checking the very considerable clearance.

To complete the story of reconnaissance soil mapping, the terms of reference of this account must be extended past independence and outside the British Colonial Service. The Netherlands government has long been a leader in foreign aid, and the most generous of all in matters of soils, based on no less than five institutions: the Royal Tropical Institute, the University of Wageningen, ITC, ISRIC, and STIBOKA. With its cadre of highly trained surveyors the Netherlands responded to a request for aid. A fact-finding mission was sent in 1970 and a bilateral agreement signed two years later. Intended as assistance in the short to medium term, it was extended for many years; a report in 1990 refers to 'donor fatigue', but aid on a reduced scale was continued well after that date.

There were two components to the Kenya Soil Survey Project, mapping and training. A national reconnaissance survey was started in 1972, initially staffed by Dutch surveyors, headed by Wim Sombroek. Field soil survey was carried out at 1:100 000 for the high and medium potential parts of the country, and 1:250 000 for the semi-arid and arid zones. By this time it had been realised that variations in climate had as much or more effect on agricultural potential as soil differences, and a survey of agroclimatic zones, headed by H.M.H. Braun, was conducted in parallel. Climate and soil information were combined in a land evaluation study, showing agricultural potential. The work was brought to a conclusion by production in 1980 of the *Exploratory Soil Map and Agro-Climatic Zone Map of Kenya*, containing four desktop-sized maps at 1:1 000 000 scale:

- Exploratory soil map.
- Agroclimatic zones.
- An overlay of the two above, for use in land evaluation.
- The information base: areas covered by surveys at various scales, and outside of these, the traverse routes and actual sites of soil observations.

Kenya occupies 580 000 km^2, making it the largest tropical Commonwealth country with full coverage by soil survey. Despite its self-effacing title 'exploratory', the soil map carries a formidable amount of detail, vastly in excess of its predecessors. The legend is found not on the map but occupying twenty-two pages in an accompanying booklet. Mapping units consist of one or more letters indicating landform group (e.g. U: uplands, M: middle-level), a further letter for the geology or parent material (e.g. G: granites, N: gneisses), followed by numbered soil units, in descriptive terms followed by the soil classes according to the FAO system, thus:

> UmGl0: Middle-level uplands, developed on granites: well drained, deep to very deep brown to dark brown, friable, sandy clay to clay (ferralo-orthic ACRISOLS)

Many units are longer and more complex than this. A legend like this displays the thoroughness characteristic of Dutch scientific work, although it is somewhat formidable for non-specialists.

The map of agroclimatic zones is based primarily on moisture availability, the ratio between rainfall and potential evaporation, with additional shading showing temperature. Developed from an earlier Kenya classification system, there are just seven moisture zones, from I: humid to VII: very arid. This makes for maps that are simple and easy to interpret, and in fact these zones have been widely used in agricultural planning and development.

The second element of the aid project, staff training, was successful, leaving a body of Kenyan soil surveyors, trained to high standards and inducted into field survey through working alongside the Dutch. This has led to a viable national soil survey, wholly locally staffed, the first director being Fred Muchena. Suffering from the problem common in developing countries of an insufficient operating budget, it has recently adopted the practice (now found in the UK) of demand-driven surveys, paid for by clients.

We cannot leave this project without saying something of its leader. Wim Sombroek was a great character in tropical soil science. After early survey work in Amazonia and Northern Nigeria he became the first manager of the Kenya project. He then went on to a distinguished international career, as Director of ISRIC, Secretary General of the International Society of Soil Science, and Director of the Land and Water Development Division of FAO. This was the more remarkable in that he was not a natural administrator, being more interested in the scientific work that was being done. Thus at FAO he personally wrote an account of the effects of global warming on agriculture, where most people in such a position would have delegated this to a consultant. He was a powerful man with a large moustache and an authoritative manner, helping young scientists as if he were a 'soils uncle'. At the closing session of a conference in West Africa in 1975 the organizer, Henry Obeng, called for a resolution that more research was needed on laterite (this being Obeng's speciality). Wim objected that the meeting had not once discussed this. When the recommendation was not withdrawn, he said in his pungent tones, 'In that case please record that I didn't vote for it.' Another non-British contributor to Kenyan soil survey was Stein Bie, a young Norwegian seconded to work with John Makin on a series of detailed surveys 1966-67. Stein was to become the first FAO authority on geographical information systems and quantitative methods, subsequently applying his expertise more widely as one of his country's leading computer consultants. We owe these contributions to a narrow escape: in his first Kenyan survey he came within a whisker of stepping on a sleeping crocodile.

Tanzania (Tanganyika and Zanzibar)

From 1947 to 1968 two surveyors, both with the surname Anderson although unrelated, spanned a total of twenty-one years in Tanzania. The younger, Gordon, was sent by the Colonial Office to understudy the older, Brian, during his first six months, after which a four-year overlap, 1958-1962, ensued. This lent an outstanding continuity to knowledge of the land resources of the country.

Brian Anderson arrived in 1947, one of the earliest post-war appointments. Initially working on fertilizer experiments, he turned his attention to survey in 1951. There was Milne's pre-war reconnaissance map of East Africa as a basis, and the Tanzania part of this was never systematically re-surveyed. Instead, knowledge was built up on the basis of localized surveys over the years, which his successor Gordon Anderson put together for the *National Atlas of Tanzania* in 1967.

An early pressing task was on hand, to set right the damage done by the lack of adequate survey and planning in the East African Groundnuts Scheme. Backed by the Overseas Food Corporation, surveys were made of two of the three areas where the Scheme had begun operations, Kongwa and Nachingwea. Kongwa lies in the nearly level, dry, central part of Tanzania, contrasting with the more varied terrain around its borders. With a rainfall of 500 mm it is not desert but carries a thorny scrub, low in grazing value and with a monotonous uniformity. The soils are 'dead' plateau soils, structureless and hard, every mineral they once possessed having long since been weathered and leached out. Nachingwea in Southern Province is more humid but suffered from tsetse infestation.

It was decided to undertake detailed surveys of sample areas, 40 km^2 at Kongwa and 52 km^2 at Nachingwea, seeking an in-depth knowledge of the management problems of these soils. Rather than depend on local staff, however, the Board of the Overseas Food Corporation decided to invite the maverick Cecil Charter, Director of Soil and Land Use Surveys in Ghana. Charter spent a month in October 1950, doing a field survey of Nachingwea. Besides his enormous energy, the report demonstrates his trenchant and sometimes undiplomatic style of writing:

> Whilst members of the Scientific Department have put in a great deal of hard and enthusiastic work one cannot help feeling that their interests have covered too wide a field ... Some of the more fundamental tasks that should have been accomplished earlier appear to have been neglected or overlooked, notably the identification of soils, determination of their distribution and assessment of their value, and ... the handling of the difficult and inherently structureless soils of the uplands on which the majority of the groundnuts will be planted ... Direction of a research organization needs wide experience and mature judgement.

Charter then demonstrates what was 'neglected' by a comprehensive survey: climate and geology, the grouping, characteristics and distribution of soils, the difficulties of large-scale mechanized cultivation, and (following earlier planning disasters) 'the experimental character of the groundnuts scheme.' The quartzite stone lines, reinforced by iron concretions, are described. 'The Scientific Department claim to have detected a B horizon heavier in texture ... I have seen no evidence to support this contention.' The basic finding is that only 32% of the soils are suitable for mechanized groundnut cultivation, and these occur as scattered parcels. Since most previous commercial operations had concerned perennial crops:

> Considerable difficulty will be experienced in finding men capable of developing the necessary skill in carrying out arable agriculture in the hot tropics ... It seems obvious to me, however, that the scheme cannot continue along its present lines with any prospect of success whatsoever.

The Tanzania Department doubtless derived much benefit from this advice, but one suspects that they might not wish to invite him again.

Brian Anderson was subsequently seconded to an FAO team making a survey for possible development of irrigation in the Rufiji basin, which covers about a third of the country. Making extensive use of air photography, he surveyed three areas for possible large-scale irrigation schemes. 'The leader of the scheme was a bit of a megalomaniac' he writes, 'and adopted a million acres (0.4 million ha) as his unit of measurement. Not surprisingly, nothing was done.' Other reasons were that the Rufiji, although rising in well-watered highlands, is not one of Africa's great rivers, and also it was being realised that the large-scale dam-and-canal type of irrigation was not an efficient form of investment. It is not entirely true that nothing was done, one outcome being the Kilombero Sugar Scheme. After leaving Tanzania Brian Anderson joined Huntings and remained with them until his retirement.

Gordon Anderson had studied soil science at Newcastle. His first experience of soil survey was the methodical but very slow British kind, field by field auger bores in the North Yorkshire Moors, before learning what matters in the tropics from his predecessor. He worked for a short time on the Groundnuts Scheme area of Nachingwea, then surveyed both forested and settled areas of the Usambara Foothills (which Geoffrey Milne had known well) with a view to cocoa expansion.

Gordon's major work was a semi-detailed survey of the southern and eastern footslopes of Mount Kilimanjaro, a zone of dissected volcanic lavas. It belongs to those areas of Africa, relatively small in extent, on which fertility is retained year after year — in modern terms, agriculture is sustainable. Sociologists are apt to attribute this to the ingenuity of the farmers, hinting that the same could be done elsewhere. Certainly the mixed cropping of coffee, bananas and vegetables, an indigenous system of agroforestry, is part of the reason for this. Fertility is only sustained, however, because rock minerals remain within the weathering zone, which leads to an ongoing release of nutrients. The survey led to maps of soils, vegetation, and land-use potential. A typical mapping unit is, 'Deep, fertile clay loams on slopes of 5-20° under rainfall of 1000-1500 mm: coffee, bananas, citrus, legume-rich pastures, vegetables.' Later, when agroforestry was first recognized as a science and a survey of indigenous agroforestry systems carried out, the Chagga system of mixed cropping on Kilimanjaro was the first to be described, in 1984, by Erick Fernandes.

The Kilimanjaro study took up about one fifth of Gordon Anderson's time, the rest being spent on detailed surveys. The titles of these indicate their practical orientation, for example a survey of Maramba 'with particular view to cocoa potential,' and of Lolkisale 'particularly with a view to Navy Bean production.' Other studies were directed at wildlife management in Tanzania's national parks: soils in relation to grassland types and grazing patterns on the Serengeti Plains, and vegetation and its utilization by wildlife in the magnificent Ngorongoro Crater. There was also much work on the nutrient requirements of ten annual and perennial crops, carried out in association with three specialist commodity research stations, particularly on responses to potassium fertilizer.

After Tanzania's independence in 1961, Gordon Anderson continued in post for a further seven years. He later helped to set up a new Faculty of Agriculture at the University of Zambia (1969-71), before moving to Makerere University, Uganda. He was responsible for a concept which was ahead of its time, 'Grow the soils to grow the crops.' The idea originated from his teacher at Newcastle, Edward Crompton, who had applied it to grasslands in northern England. Gordon took this concept and applied it to the tropics, collecting examples of sustainable cropping systems in Africa. As a chemist he well knew the role of nutrients, and where possible fertilizers, but as a field scientist he appreciated also the need for maintaining physical structure and biological activity. A soil requires protection from erosion, feeding with both chemical fertilizers and organic manures, encouraging deep and prolific root systems, and regular examination (through profile pits) as a basis for diagnosis and remedial measures. Growing a soil 'may be likened to a child growing up and needing protection, good food, and an environment ... in which it can put down roots into the community ... able to withstand the hazards and setbacks of life.' In the 1980s this approach was to be developed through the Tropical Soil Biology and Fertility programme. It is also applied, albeit downplaying the need for fertilizers, in the modern concept of conservation agriculture (Chapter 13).

Recollections

Anthony Mitchell was Project Manager of the Tabora Rural Integrated Development Project, heading a multidisciplinary team of eight staff. He describes an intercultural misunderstanding. When this project started in 1978, it was interrupted by the war with Uganda which led to the overthrow of Idi Amin. Tabora airfield became an important base, and security was tight. Some time after the war was ended, they held a Christmas party for the children. Among the team members was a keen amateur pilot, and someone thought it would be a great idea for Father Christmas (Jim Mansfield) to arrive by air. The pilot 'buzzed' the houses and headed for the airfield, and the children piled into Land Rovers to meet him, only to return looking forlorn, saying, 'Father Christmas has been arrested.' It seems that having landed, the pilot felt he must have a photo of Father Christmas by the plane. Even at the best of times, taking photographs at airports in developing countries is strongly inadvisable, and when this strange figure arrived, his face covered in cotton wool, it was too much for the security guard, who arrested them at gunpoint. Colleagues went to explain, and were asked, 'Why is that man disguised? He must be a spy.' Showing a photo taken in the plane (it was a Polaroid camera) they were eventually able to explain, and then the police all wanted photos of themselves — in the Police Station, normally another photographic no-no. So all ended amicably.

Mitchell also reports a minor but unusual achievement in Tanzania. When working on ujamaa (freedom) village reorganization his team, wishing to show local farmers their objective, adopted as a logo to paint on the doors of their vehicles, *Matumizi Bora ya Ardhi*, or 'Best Uses of Land'. When he returned in 1985 the Land Use Planning Division of the Ministry of Agriculture was called *Idara ya Matumizi Bora ya Ardhi*.

Land information systems

Among developing countries, Tanzania was one of the first to set up land information systems, digitized databases in which soils information was were combined with data on climate, water resources, vegetation and land use. In 1967 a research unit, the Bureau of Resource Assessment and Land Use Planning (BRALUP), was set up in 1967 at the University of Dar-es-Salaam. BRALUP was the brainchild of an English geographer, Len Berry, at the time Dean of Arts and Social Science. He observed a not uncommon phenomenon, that economists were cornering the market in giving advice to government, through an Economic Research and Development Bureau. The concept of BRALUP was to focus on rural areas from a geographical development perspective. The idea took hold, Norwegian aid contributed a six-storey building, the Rockefeller Foundation supported a director, and by 1972 it had grown to a staff of fourteen. An American, Bob Kates, was the first director, then Berry, and in 1972 a Tanzanian, Adolpho Mascarlin, took over and it continued to operate with mainly Tanzanian staff. A major project of the early years was a study of soil erosion and sedimentation rates, headed by a Norwegian geomorphologist, Anders Rapp, whose previous career had been spent studying glaciated landscapes in the Arctic. This is one of the few cases in which erosion was actually measured in the field, as distinct from being modelled.

In the early 1990s much of the existing information on land use and environmental degradation was out of date. In the first instance revision was needed for a Forest Resources Management Project. BRALUP was reborn as the Tanzanian Natural Resources Information Centre, located at the University of Dar es Salaam. By means of a joint project with Cranfield University, UK, funded by the World Bank, a Tanzanian Natural Resources Information System (inevitably TANRIS) was set up. Besides its original need for forest management, this has been applied to work on water resources, vegetation changes, agro-economic zones, and even population changes.

Zanzibar

Zanzibar's period as an independent country was the shortest in the Commonwealth. Given the status of a British Protectorate in 1890, it was granted independence as a constitutional monarchy in 1963; the following year the Sultan was overthrown and it became part of the United Republic of Tanzania. Too small to support a soil survey, this was conducted by four-way collaboration between the Tanzania and Zanzibar Departments of Agriculture, the Macaulay Institute of Soil Research, Aberdeen, and the Commonwealth Bureau of Soil Science.

W. E. Calton, Soil Chemist in Dar es Salaam, made a soils reconnaissance of the two main islands, Zanzibar and Pemba, in 1949, subsequently expanded by a two-month survey of Pemba Island. The outcome is a comprehensive description with soil maps and chemical data; agricultural potential and problems were expanded through collaboration with a local agricultural officer. Notably for the time, although the soils are described in scientific terms, the mapping units are based on locally recognized soil types (*mchanga, kinongo, utasi*, etc.), a practice nowadays dignified by the name ethnopedology. 'Discussions with the native peoples during the survey brought to light many of the soil names ... [which] are adopted tentatively as pedological terms.'

Thus in the three main countries covered by Geoffrey Milne's pre-war survey, and which were for a time united by the East African Agriculture and Forestry Research Organization (EAAFRO), later work followed different lines. Uganda was the most successful, with national coverage by soil and ecological surveys accomplished in a single effort. In Kenya there were two phases, local surveys 1958-1970 with an interim compilation at 1:3 000 000 scale in 1959, followed by a systematic national survey at 1:1 000 000, financed and largely staffed by the Dutch, 1970-1982. Tanzania followed the line of local, purpose-directed studies, but apart from a compilation at 1:3 000 000 for the National Atlas in 1967, never undertook a systematic national survey. Milne would have been gratified that his calls for more detailed work had been heeded, and surely also proud at what he had accomplished single-handed.

Somalia (British Somaliland)

British Somaliland belonged to the Commonwealth until 1960, when it joined with French and Italian Somaliland as the independent state of Somalia. Prior to this date, there are no records of surveys. A number were conducted by Huntings and Macdonald for irrigation and livestock projects during the 1970s. The most substantial work came in 1985, when the Land Resources Division conducted surveys for a National Tsetse and Trypano-somiasis Control Project, including maps of land use, land capability, and livestock distribution and seasonal movements, although this was for an area formerly part of Italian Somalia.

Sudan

From the beginning of the twentieth century Anglo-Egyptian Sudan was a jointly administered condominium. The political niceties of this situation need not concern us. The annual staff lists are in the same format as those of other colonies and show that in the Department of Agriculture at least, the expatriate staff were almost exclusively British. In 1956 Sudan became the earliest country in the British Commonwealth to achieve independence.

In 2011 South Sudan, mainly the savanna area, became an independent country, but has since suffered from severe civil disturbance.

With an area of 2.5 million square kilometres, over ten times the size of the United Kingdom, Sudan is Africa's largest country. A national reconnaissance survey was out of the question, nor would it have served the country's needs. For half a century the history of soil science in Sudan was largely that of the Gezira, moving into other areas from 1952. The Gezira, Arabic for 'island' or 'peninsula', is the area lying between the White and Blue Nile Rivers upstream from their confluence at Khartoum. It is a featureless clay plain, roughly a triangle of base 120 km and length 240 km. What makes it special is that the clay is derived from the volcanic mountains of Ethiopia, conferring on it the high fertility of soils derived from dark, basic rocks.

An early air photograph: the Gezira Research Farm, Sudan, 1929

After a quarter-century of discussions and planning, the Sennar Dam on the Blue Nile was completed in 1925, from which sprang a canal irrigation scheme. The Gezira Scheme was based on a partnership between the government, the commercially-based Sudan Plantations Syndicate, and the farmers, the latter holding their land as tenants. Long-staple Egyptian cotton was the major cash crop, originally destined for Lancashire mills. Egypt was not unnaturally concerned at the prospect of Sudan taking its irrigation water. This problem was settled by the Nile Waters Agreement

1929, a consensus easier to reach when Sir Murdoch Macdonald, founder of the engineering firm which bears his name, was not only prominent in construction of the Gezira but also Irrigation Adviser to the Egyptian government. During the post-Second World War period when not a few development projects were disastrous failures, the Gezira came to be cited as the prime example of a successful scheme, the result of which came from discussion, research and surveys, planning over many years, and good management. Those interested can compare the approving but somewhat paternalistic account by Arthur Gaitskill, *Gezira* (1959), with the social-ist-orientated views of Tony Barnett in *The Gezira Scheme: an illusion of development* (1977).

The research and planning of irrigation projects depends on cooperation between engineers, soil scientists and agriculturalists. With land slopes of between 1 in 5000 and 1 in 10 000, the layout of canals called for highly precise surveys.

Large savings in costs arose because the impermeable clay soils did not require canals to be lined. Agricultural research began at Khartoum, but from 1918 was based on the Gezira Research Farm at Wad Medani. A chemist, W. Beam, began laboratory examination of soils in 1904, and by 1911 had produced a preliminary report on Gezira soils, based on five east-west transects. Little known, it could be argued that Beam was the earliest British tropical soil surveyor, before the pioneers described in Chapter 2. Certainly Mohammed Abdul El A'al, appointed Chemist in 1929, was the first locally-born soil surveyor in Africa by many years. He participated in a supervisory capacity in many field surveys, and was still in service in 1955. Beam was succeeded by A.F. Joseph in 1919, after which Sudan was to provide the first, or early, posts for a remarkable succession of soil scient-ists: Frederick Martin (1919-1924), Herbert Greene (1924-1945), Tom Jewitt (1936-1955), Alun Jones (1945-1953), and Colin Mitchell (1952-1955).

In the absence of identifiable landforms, soil survey on the depositional plains in Sudan took the form of grid survey. In the early years, 'the standard pit adopted was 10 ft deep, 12 ft long and 3 ft wide (3 x 4 x 1 m), with 1 ft steps cut to the bottom', a standard of luxury which latter-day surveyors would envy. The original maps for the Gezira scheme lie deep in the archives as *Sudan Gezira Reports*. It is apparent from the detail shown that the sampling pits must be numbered in hundreds of thousands. Linked with field surveys were extensive laboratory investigations at the Gezira Station. As regards soil chemistry, what mattered for irrigated agriculture were the percentage of salts and the degree to which these had saturated the clay complex. The most innovative work was the pioneering studies in soil physics by Herbert Greene.

In 1952-53, surveys were carried out for the Manaqil Extension to the Gezira scheme, almost as large as the original area. Both before and after independence there were also pre-investment surveys, many on a large scale, in other parts of the country: Jebel Marra in the west, a Savanna Development Project, and for the Nile Waters Agreement. Two resettlement projects, for people displaced by the construction of the Aswan High Dam, and for settlement of refugees from Eritrea, now appear ironic in the light of the large numbers of Sudanese refugees displaced by recent civil disturbances. These large projects were sponsored by UNDP, other international bodies, and the British Overseas Development Agency, and conducted through partnerships between consultant firms, notably Macdonald for engineering studies and Huntings for soils and agriculture. Preparatory surveys for the Jebel Marra Rural Development Project in the west of the country were among the most comprehensive ever undertaken, beginning with land and water resource surveys 1968-69 and ending with a Final Report in 1979, to be followed by annual reports of the executed project. Among surveyors taking part were Anthony Mitchell (1962-1964), Paul White (1968-1969) and Maurice Purnell (1973-1977).

When Sudan became independent in 1956, this did not interrupt the continuing flow of surveys, largely directed at irrigation potential. In 1957-63 Huntings surveyed 30 000 km^2 of the Jebel Marra in Darfur Province. Further donor-funded surveys included Kordofan Province in 1963 and the Roseires soil survey of 1967. Irrigation surveys continued into the 1990, and work was extended to surveys for pump irrigation.

With more than half a century of soil surveys, largely for irrigation potential, Sudan has the largest number of entries in the World Soil Survey Archive and Catalogue (WOSSAC), at over 1800 items. These surveys led to a vast extension of irrigated agriculture on which the economy of the country, and the livelihood of its people, depend.

A noteworthy feature is the way that so many of Sudan's scientists went on to distinguished careers in other countries. Frederick Martin's initiative in starting soil survey in Sierra Leone has been described in Chapter 2. Herbert Greene was to become Tropical Soils Adviser to the UK government. Tom Jewitt became a leading soil scientist with Huntings. Alun Jones went on to take charge of soil survey in the West Indies. Colin Mitchell worked for Huntings before a career in the Department of Geography at Reading University, where he became a specialist in air-photograph interpretation. Maurice Purnell, after four years in Sudan on an FAO project, became a stalwart of the Soils Service at FAO's headquarters in Rome, taking responsibility for land evaluation.

Why should Sudan, with its vast areas of almost featureless plains and limited range of soils, have been such a fertile proving ground and inspiration? The answer may lie in institutional features: an early appreciation of the role of research, construction of research facilities, a close link with extension services, and continuity. Early work was financed by the government subvention from revenues of the Gezira, a scheme on which the need for research, both before the project and ongoing during its implementation, was always recognized. Later projects benefited from international financing and, at that time, favourable conditions for investment. This led to an institutional environment which enabled scientists to get on with their research, build up experience over many years, and pass it down to their successors.

Recollections

Colin Mitchell, describing a 1952-53 survey in Northern Province linked with the Nile Waters Agreement, gives an idea of what survey was like in Sudanese conditions:

> The team consisted of Mohammed El A'al (now in his 23rd year of service), another Sudanese who booked the information I called out from the pits, four field assistants, ten drivers and 40 labourers. There was a fleet of vehicles consisting of four Bedford pick-ups, four Commer lorries and one Land Rover. These first had to be taken across desert country. I soon learned the method employed by Sudanese drivers in soft sand: go in low gear at high speed and don't stop or declutch until you reach solid ground. The Survey Department had preceded us and laid out a series of lines 10 km

apart at right angles to the Nile, with heights above the river marked on concrete posts like Belisha beacons.

The procedure was to send out the Commers in the morning, with the driver dropping two labourers at sites 1 km apart, where they would dig a 2-metre pit like a grave. Thus they prepared 20 pits for inspection the following day. El A'al and I went out separately in our Bedfords and each described and sampled ten pits. The large number of samples was sent back to Wad Medani, imposing such a burden on the chemical section that they mixed the top metre and gave us only a single salinity and pH reading from each pit. We had not yet adopted the now recognized method of using geological and topographic information to identify typical sites, which were studied in detail, and then relate other sites to these. But at about this time it was realised that grid survey was costly and inefficient.

The next summer I was sent to survey Yambio Experimental Farm in the extreme southwest of the country. The former District Commissioner there, 'Tiger' Wild, had the idea of improving the local diet by inducing his officials to carry petrol tins full of mango seeds and throw these out onto the roadside. The result was that most roads around Yambio were lined with huge mango trees.

The mention of petrol tins demands a digression. They were, I think, the invention of the Shell company. Each held 5 gallons (23 litres) and were made of tinplate, with a small carrying handle and plug. These were the usual means of distributing petrol around the country, but their re-uses were endless: transporters of water and foodstuffs, waste paper baskets, tied in sequence to form scoops (saqias) on water wheels, cut up and joined to form roofs and walls. Fashioned with a pair of snips, they served for vehicle and domestic repairs, and appeared in markets in practical and decorative guises such as dustpans, parts for pressure lamps, and picture frames. Mitchell continues:

> Both here and in the north we frequently found snakes that had fallen into pits dug the night before. On one occasion one of the Bedfords would not start, and on opening the bonnet we found the reason was a snake which had wound itself around the radiator, preventing the fan from working. In the north, leaves also fell into the pits, and on one occasion I became conscious of a large snake

under my feet. The Zande assistants dragged me out, threw down stones until they killed the snake, then draped it artistically beside the pit. When I enquired the reason they looked reluctant, until one of the assistants explained, 'You see, Sir, they want to save it to eat.' In that tsetse-infested land any meat was valued.

Alun Jones was asked to look at a remote area in the south-east of the country to assess its suitability for mechanized cultivation of sorghum. It was a broad expanse of open plain with few landmarks, so anything that might help find where you were was a bonus. The maps of the time showed 'Murray's Trace', a survey line cut in the early part of the century by a military engineer, presumably named Murray:

> Accordingly, Fred Hoyle and I set out one afternoon in a Bedford truck with driver to see if we could find Murray's Trace. We rapidly found that the grass that was head high around us was springing up to the rear and covering our tracks, and in no time we were completely lost. It was elephant and lion country and dusk was approaching when the animals could be expected to head for the river to drink. We tried doing a 360-degree turn in the hope of finding our outbound track but to no avail. By this time the moon had risen and we luckily stumbled on a Dinka encampment and asked for directions. Pointing these out, they said we might just reach Renk, our base, by the time the moon had gone from the sky — and they were right, too!

Herbert Greene (1897-1964)

One of Sudan's 'alumni' deserves further notice. Herbert Greene belonged to the distinguished generation who served in both world wars. His university career at St Andrews was interrupted by the First World War, where he fought in Flanders with the artillery and was awarded the Military Cross and bar. In the Second World War, in his forties, he served in the Abyssinian campaign as Adjutant in the Sudan Defence Force.

Graduating in chemistry with distinction, he continued at St Andrews for three years doing research. In 1924 he was appointed Chemist to the Sudan, where his work over the next twenty-one years did much to contribute to the success of the Gezira scheme. In particular, he realised the importance of soil physical studies, primarily into the retention and movement of water and salts. The general reader may not be stirred by the title of his 1928 paper, *A soil profile in the eastern Gezira*, but to the committed soil scientist it is a classic.

For four years after leaving Sudan Greene worked for FAO, before joining the staff of Rothamsted Experimental Station as Tropical Soils Adviser, a post financed by the Colonial Office. From this base he embarked on extensive travels having a mission, as he saw it, to bring recent advances in soil science to the attention of chemists and surveyors in the colonies who were cut off from academic contacts.

He would also publish summaries of new work; a 1961 paper, *Some recent work on soils of the humid tropics*, although only three pages long is a *tour de force*. When work had been published demonstrating the relation between iron oxide reduction and phosphate fixation, Greene realised its importance but also the fact that the original paper was somewhat obscure, and explained its significance in his 1962 Presidential Address to the British Society of Soil Science.

Greene continued to undertake strenuous tropical journeys despite fading health. 'Unfailingly friendly, courteous, and helpful,' reads his obituary, 'his wise advice was continually sought, and never in vain, but especially note-worthy was the encouragement of stimulation he provided for junior soil scientists working in remote places.' I was one of these, visited by Greene in Malawi in 1960, where his message concerned the recently-discovered variable-charge clays and their relevance to soil fertility. Greene died in Lagos in 1964 whilst on a technical mission to West Africa.

5

WEST AFRICA

Ghana (Gold Coast)

THE PRESENCE OF C.F. CHARTER at the cocoa Research Institute during the last years of the war gave the Ghana a flying start in soils research. In 1948 he became founding Director, and sole staff member, of a Soil Survey Division in the Department of Agriculture. Three years later this became the Department of Soil and Land Use Survey. A favourable attitude towards aid by the Colonial Office, combined with Charter's persuasiveness, led to a rapid expansion of his Department. After an initial temporary site in the Botanical Gardens, Aburi, it moved to purpose-built accommodation at Kumasi in 1953: offices, library, soil museum, herbarium laboratory, photographic dark room, carpenters' and blacksmiths' shops, and its own electricity generators. A fleet of twenty Land Rovers and three-ton lorries was maintained. The first five soil survey officers, together with an analyst, were in post by 1951, rising by 1954 to ten surveyors, two soil analysts, and a compilation officer. With this kind of set-up, conditions were in place for a five-year period, 1951-1955, of great productivity.

Recruitment began with Hugh Brammer, Stanislav (Stan) Radwanski and Ron Philp in 1951. Alan (Sandy) Crosbie came the following year, Maurice Purnell and Alan Mould in 1953, and Peter Ahn, Alan Stobbs and Helen Brash in 1954. Charter displayed two preferences in his recruitment, for geographers and unmarried personnel. He selectively recruited geographers, mainly from Cambridge and Oxford, in preference to the chemistry back-ground of many early soil surveyors; this preference might have stemmed from his favourable experience with Brammer. He even set up an establish-ment post of senior geographer, but this was never filled. Although a former married man himself, he is reported as saying, 'One soil surveyor married equals half a surveyor.' This was not entirely true; Crosbie, as he puts it, 'appears to have slipped through,' being already married on recruitment. Brammer and Purnell remained unmarried all their lives, Radwanski married whilst in post, the others only after leaving Ghana. More about Radwanski's life can be found in Chapter 6.

As Peter Ahn expressed the situation, 'I went to the Gold Coast in January 1954, by boat of course, and worked with about ten other recruits of similar age and lack of experience.' Their activities rapidly got under way: a national reconnaissance survey and detailed local surveys, allied to research, both theoretical and applied. For reconnaissance work each surveyor took lead responsibility for an area: Brammer for the Accra Plains, Ahn and Radwanski respectively the Lower and Upper Tano Basins, Crosbie the area near Kumasi, Mould the south-east, and Stobbs the north. It was intended that each of these should result in maps and memoirs but the exigencies following independence interrupted this process and only two regions were published as independent reports. Ahn's *Soils of the Lower Tano Basin* is a 270-page memoir, with numerous diagrams and photographs, and three maps at 1:250 000: Soil Associations, Land Use, and Traverses and Sample Strips, this last showing locations of the detailed sample areas within each Association called for by Charter's method. The Soil Associations taken as the basis for mapping are made up of Soil Series of which seventy-two are described, for every one of which there is a substantial account of agricultural value and requirements for soil management. The maps in Brammer's *Soils of the Accra Plains* are at 1:125 000, usually considered a semi-detailed rather than a reconnaissance scale. The northern part of the country was later to be covered by a six-volume FAO survey completed in 1968.

An unpublished survey of the Upper Tano Basin, covering 9 200 km^2, has recently come to light. Stan Radwanski completed the field survey of this area in 1952-54, and during the remainder of his time in Ghana compiled the report and maps. Problems arising during the run-up to independence meant that this was never published, so to prevent it being lost he took it with him. On his death in 1993 his papers passed to his widow, Estela Radwanska, who gave them to the author in 2006. They have been deposited with the World Soil Survey Archive (WOSSAC) at Cranfield University, UK.

Radwanski's report is complete in draft: 150 typescript pages, many tables, about 100 soil analyses, and seven diagrams of soil catenas. There are no fewer than nineteen maps, neatly drawn in coloured inks and crayon shading ready for the draftsman; these include Landforms, Soils, Vegetation, Land Use, and the routes of his field traverses. It is possible that other survey reports from the Colonial period survive in government offices, yellowed and ravaged by termites.

A Technical Advisory Committee for the Development of Agriculture and Natural Resources had been set up in 1953, and at its first meeting Charter proposed that a textbook on agriculture in Ghana should be prepared, for use by government officers, university teachers and research staff. Preparation was interrupted by independence, but resumed in 1958 when editorship was placed in the hands of J.B. Wills, Compilation Officer of the Soil Survey. *Agriculture and Land Use in Ghana* (1962) is one of those massive illustrated works which the economics of publication seemed to allow at the time. Five hundred quarto pages long, with fifty-nine photographic plates and four maps, it weighs two kilogrammes. The twenty-six chapters, some co-authored, cover every aspect from natural resources to crops and, unusually for the time, social factors. There are also accounts of swollen shoot disease, capsids, fungi, black pod, the ring-barking beetle, shot-hole and pod-husk borers — it is a wonder that cocoa production was possible at all. Natural resources are comprehensively covered, including climate, geology, rural water supplies, geomorphology, soil types, fertility, erosion, vegetation, forests and fauna, with elephants bringing up the rear. Geology, vegetation and soils are accompanied by back-pocket maps at 1:2 000 000 scale, based on Charter's oxysol-ochrosol classification. Such clear and comprehensive compilations of knowledge are rarely found nowadays.

In common with many other countries, the soil survey staff by no means spent all their time on reconnaissance mapping. The 1960 Publications List gives forty-eight Technical Reports, listed as 'Usually thought to be primarily of local interest only', produced in a steady stream from 1954 to 1960. Some were interim accounts, but many were intended to evaluate new crop or development proposals, or management of forest reserves, and to place these on a sound basis. Maurice Purnell worked only on this kind of survey, and one title of his can serve to illustrate its practical nature, *Report on the detailed soil survey of the proposed oil palm plantations at Pretsia* (1960).

For Charter, the objective of soil surveys was always practical. Around the walls of the Soil and Land Use Survey he had a number of framed notices, one of which was, 'The purpose of this office is to produce maps as tools for agricultural development, not pretty-coloured pieces of paper.'

In 1962, after eight years in soil survey, Peter Ahn left to spend nineteen years at the University of Ghana, the latter part as Professor and Head of the Department of Soil Science. Having acquired a knowledge of soils in the field, he turned his attention to plant nutrients, manures, fertilizers and soil factors affecting cropping systems. He moved on to Kenya and the Cote d'Ivoire (becoming one of the few British scientists fluent in French), before becoming Coordinator of the Vertisol Network for the International Board of Soils Research and Management (IBSRAM). Ahn's extensive knowledge of both the pedology and management is set out in *West African Soils* (1970).

Most expatriate staff left the country when their contracts expired in or soon after 1957; the Annual Report for that year records, 'There was no response to recruiting advertisements during the period.' Brammer, Purnell and Radwanski remained in soil survey, whilst Crosbie became Professor of Geography at Edinburgh University.

Meanwhile the education of Ghanaians was proceeding apace, and by 1959 the Soil Survey consisted of three expatriates and four local surveyors. Victor Adu was the first Ghanaian to be employed by Charter in 1945, as Senior Technical Officer. He went on scholarship to the Macaulay Institute, Aberdeen in 1953, was subsequently promoted to professional grade, and worked as soil surveyor until he retired in 1984, a fine example of personal development and sustained service.

Henry Obeng preceded Adu at professional grade, taking up post in 1956. He graduated BSc agronomy and MSc soil fertility at Iowa State College, USA, his internet autobiography stating, 'This was in furtherance of the ardent desire of the Leader of government Business, Dr Kwame Nkrumah, to quickly implement the Africanisation of all professional grades of the civil service.' By 1959 there were two substantial detailed surveys in his name. His own account states that, 'During this period he was mostly in the field either actively conducting soil and land classification or inspecting soil survey projects in every region of the country', and at the same time he conducted studies which led to a PhD on iron-pan soils in 1970. In the same year, Obeng succeeded Brammer as Director of the Soil and Land Use Survey, later the Soil Research Institute. For a time he was one of the leading internationally-recognized African soil scientists, putting him in a position to be invited onto international committees. In 1975 he organized a Commission meeting of the International Society of Soil Science, followed by a field tour, in which the normal activity of inspection of soil pits was interspersed by visits to District Commissioners, Chiefs and other dignitaries; Wim Sombroek remarked, 'Dr Obeng seems to have invented a new branch of soil science, Soil Politics.' Obeng moved into consultancy work in 1982.

Two of the earliest African soil surveyors:
Henry Obeng and VictorAdu, Ghana

Other early Ghanaian surveyors were Frank Ablorh, E.O. Asare, P.E. Danka (whose soils education was in East Germany), and Kwasi Quagraine. Asare left to become Professor of Soil Science at Kwame Nkrumah University, Kumasi, whilst Quagraine took over from Brammer as head of the Soil and Land Use Survey in 1961.

Distinguished visitor: the Duke of Edinburgh is shown soil samples
awaiting analysis, Kumasi, Ghana; Hugh Brammer on right

In 1957 Ghana became the second ex-colonial country to achieve independence, regrettably followed by a period of political instability which handicapped agricultural production, particularly of cocoa. Surveys for development projects, mostly irrigation schemes, continued in the 1970s, notably a series of reports on the irrigation potential of the Accra Plains. In 1976 the Commonwealth Development Corporation produced a report on oil palm development sites.

Ghana was an early leader in development of local staff, and soils research has continued to be active. The CSIR Soil Research Institute, based on Kumasi, currently has a scientific staff of over 25. As in many countries, survey is no longer a major part of its work, being able to draw upon the existing national coverage by reconnaissance surveys. These have been converted into a soil data base, part of a geographic information system. Notable among recent research projects is a study of soil health policy.

Whilst not a survey, Ghana was the site of a fundamental piece of research, leading to a seminal advance in knowledge. Peter Nye joined the Ghana Colonial Service in 1947 as Agricultural Chemist, and after a short spell in Nigeria became Senior Lecturer in Soil Science at the University College of Ghana, subsequently the University of Ghana. He was joined in 1954 by a young Research Fellow, Dennis Greenland. Together they looked into what was happening to the soil in shifting cultivation, why it had the potential to maintain soil fertility, and what happened when fallow periods were reduced. Field observations led to a theory of soil organic matter, dividing it into fractions with different rates of decay. The key lay in the decomposition constant, the percentage of the pool of organic matter lost by oxidation each year. This was higher under cultivation than under forest fallow. By modelling theoretical calculations to field observations they showed that the fallowing system was far less efficient in the savannas than in the forest zone, and that both systems would inevitably lead to soil degradation once the ratio of cultivation to fallow went above critical limits. The results were set out in *The Soil under Shifting Cultivation* (1960), one of the most influential and frequently cited publications in soil science.

Greenland does not recall the budget for this research project, but by modern standards it would seem miniscule. He can, however, remember his excitement at receiving his first pay envelope, expected to contain the first instalment of an annual salary of £450 per year, and finding that it was 'You owe us so much', the charge for staying the first two weeks after arrival in the College Guest House having exceeded his pay for the month.

Nye's subsequent research was primarily at Oxford University, working on the movement of soil nutrient ions. Greenland had probably the most distinguished career of any tropical soil scientist, British or other, at the Universities of Adelaide and Reading, and as Director of Research at two of the international agricultural research centres, the International Institute for Tropical Agriculture (IITA) at Ibadan, Nigeria, and the International Rice Research Centre (IRRI), Los Banos, the Philippines. He possessed an extraordinary capacity for keeping abreast of scientific developments whilst shouldering a heavy administrative burden. Both Peter Nye and Dennis Greenland became Fellows of the Royal Society.

Recollections

Sandy Crosbie is one of several correspondents who draw attention to the need to inspect soil pits for snakes before entering. 'If not, you made a standing jump which should have got into the Guinness Book of Records.' It is common practice when beginning survey in a new area to visit the local chief and village elders. Schnapps, introduced to West Africa by the Dutch, was the spirit usually employed on these occasions. One sat in a circle and a large tumbler of it was passed around, placing faith in the sterilizing power of alcohol.

Helen Brash

Charter's requirement that his soil surveyors should be unmarried cannot wholly be equated with male prejudice, for he also recruited Helen Brash. Women were not numerous in the Colonial Service, and so far as can be ascertained, she is unique among Colonial soil surveyors. Women staff were equally rare in topographic survey, geology and forestry.

Brash took the Cambridge geography tripos, selecting as final year options physical geography and surveying, the only woman to take the latter. Undergraduate expeditions are a wonderful preparation for jobs requiring fieldwork and she went on two, one by getting herself invited onto a Nottingham University glaciological expedition to Iceland. (I was to follow a similar route the following year, on an expedition to a Norwegian glacier.) 'I found most doors closed to me, including the Ordnance Survey, because I was a woman. I applied for a Nature Conservancy scholarship, but was perhaps unwise to go to the interview wearing a smart suit and a hat!' No jobs of a field nature being available, she started a Diploma of Education. During the first term of this, Charter came to the Geography Department seeking recruits, and Brash's supervisor, Bruce Sparks, introduced her to him. 'I knew precious little about soils, but 'survey' sounded good.' Charter told her to apply, said the Crown Agents would interview her but not to worry about that, she would get the job. On that occasion she did not wear a hat.

The only woman soil surveyor in the British Colonial
Service, Helen Brash, with C. F. Charter, Ghana

Having accepted the job because it involved fieldwork, on reaching Ghana
she found herself appointed to the post of Soil Correlation Officer at
headquarters in Kumasi. 'Charter thought this a 'safe' job for a woman,
where he could keep his eye on me. I learnt afterwards that he had written
to my mother saying he would not send me out into the bush on my own.
You can imagine my reaction!' A car was ordered, but where other officers
were given a Standard Vanguard estate model, she was told that as a woman
she should have a saloon:

> Charter was a wonderful teacher, but my field experience on the
> first tour was confined to occasional trips. Towards the end of that
> tour he allowed me to spend some time in the north with Alan
> Stobbs and Buchanan Brand. One day we all felt in need of an
> afternoon siesta. I awoke to find a string of village women walking

past my tent, ostensibly to collect water but I was told they had taken that route because they had never seen a white woman.[1]

When the book, *Agriculture and Land Use in Ghana* was being prepared, Brash was given the office job of writing the chapter on geomorphology. However, during her second tour of service (in West Africa, because of the presumed more oppressive climate, these were only eighteen months) Charter relented and put her in charge of a regional survey. This was the Ho-Keta plains, an anomalous area of savanna woodland east of the Volta River where, because of a cold upwelling of water, the rainfall was too low to support forest. 'I had an excellent team of Gold Coast staff, led by A.K. Akutor, Assistant Soil Survey Officer, who had a wealth of field experience and was extremely valuable.' Survey was interrupted by her secondment as a returning officer for the Togoland plebiscite.

At the end of her second tour in 1957, Brash married the manager of a local subsidiary of Unilever, becoming Mrs Helen Sandison. The Survey drew the line at employing a married woman — indeed, the Colonial Service as a whole accepted very few — so she turned her hand to geography teaching, first in Kumasi, then Lagos, and finally in England. On retirement on 1994 she took an external theology degree at Oxford, thereby becoming a graduate of both Britain's older universities. Hugh Brammer describes Helen Brash as the first feminist he met. In the light of her experiences, this is not surprising.

Nigeria

Cocoa has several times provided the stimulus for land resource surveys. Before the Second World War, Frederick Hardy made it his focus of interest in the West Indies. After the war, the need for replanting following clearance of areas affected by swollen shoot virus provided the initial incentive in Ghana. In Nigeria, cocoa was also a major crop in the Western Region. The Cocoa Marketing Board, conscious of the threat posed by swollen shoot virus and realizing that new planting might be necessary, set up and initially funded the Cocoa Soil Survey, covering areas climatically suited to the crop. Happily, its investigations were broad-based and provided valuable information for a wide variety of general agricultural planning. Harry Vine was its

---

[1] My wife had the same experience in Malawi in 1958.

first Director (1951-56) and Tony Smyth the second (1956-62). Other staff were Roy Montgomery, Rowland Moss, Peter Thomas and Bill Foster.

The field teams were large: one soil surveyor, two senior and ten to fifteen junior assistants, fifty to sixty labourers and two to three drivers. Smyth describes the survey methods:

> The only air photography available was of such poor quality as to be almost useless. Roads were few and access between them extremely limited. The reconnaissance survey was conducted, therefore, by ground survey methods, using compass traverses four miles (6.4 km) apart with soil profiles at intervals of 10 chains (200 m). Trained recorders sketch-mapped the vegetation and land use in a circle around each sample point, and altitude was measured by aneroid barometer. Soil horizons were laid out in lengths of split bamboo. These methods, developed by Harry Vine after experience with Charter in the Gold Coast, proved very effective.

Local detailed surveys were carried out for immediate advisory purposes, such as on research farms. In all, 100 000 soil profiles were examined.

The Soils and Land Use of Western Nigeria, written by Smyth and Montgomery, appeared in 1962. It was the first book of its kind produced by a newly-created government printer, which, Smyth remarks, 'went to town with colour plates and hardback binding.' Presentational aspects apart, it is a classic of its kind. The greater part of the text consists of soil descriptions and relationships with environment, particularly clearly expressed and illustrated. There is a thirty-page chapter on soil use and management, in three sections: soil quality for agriculture, soil quality for engineering purposes (mainly road building, written by a government engineer), and crop management. This last starts with cocoa, summarized in descriptions of 'good, fairly good, and poor cocoa soils', which four years later was to be expanded by Smyth in an FAO monograph on cocoa soils. Soil management for coffee, citrus, oil palm and other perennial crops is then covered, followed by a key problem, the maintenance of fertility under arable farming.

The enclosures are of interest. The map shows nine Soil Associations with local names, each described in understandable terms, e.g. 'Iwo Association: sandy, greyish brown ... rock at 7 to 9 feet.' At 1:250 000 scale this is a sheet 120 by 60 cm, suitable for use in the field or to add local detail. There is a photographically-reduced version at 1:500 000, giving a 60 by 30 cm size appropriate for office work. Finally there is a 'Soil classification disk', a circular card with three layers revolving about its centre, with slits revealing, for any chosen soil series, its characteristics, area, and crop recommendations. At the time this was criticized as being over-simplified, but in retrospect it can be seen as a forerunner of the computerized soil-crop databases now in use.

From 1951 onwards, a second centre was active in the largely semi-arid region of Northern Nigeria. Alex Muir, Director of the Soil Survey of England and Wales, had been sent on a fact-finding mission and recommended creation of a Soil Survey Section within the Agricultural Research Station of Samaru. Graham Higgins, with a background of agricultural chemistry at the University of Wales, Aberystwyth, was in post as Soil Chemist, responsible for fertilizer trials and nutrient studies. He underwent a five-month soil survey training in the UK and Ghana, to become the first Head of the Survey Section. Further recruitment followed, building up a team of nine, including Kees Klinkenberg, one of several Dutch taken into the British Colonial Service. Initial work focused on assessing how representative was Nigeria's framework of experimental farms. Detailed surveys of farm sites were followed by wide-ranging road traverses, building up knowledge of areas over which experimental results from each site could be applied.

Some specific site surveys were also carried out, e.g. for sugar production and seed multiplication. By 1957 a considerable number of Nigerian support staff had been trained, and work began on reconnaissance survey at 1:100 000, making use of some poor-quality 1945 RAF air photographs. Douglas Ramsey assumed charge in 1959. In 1962 the Soil Survey Section was transferred to the newly-created Ahmadu Bello University. Higgins became its first Head of Soil Science, remaining until 1967, whilst other expatriates left on termination of their contracts. By 1965 the number of *Soil Survey Bulletins* had reached thirty, an indicator of the ongoing local detailed work. The reconnaissance work remained incomplete and, in Higgins' view, 'use of its results is still to be demonstrated.' Some of the local surveys led to early development, such as the Bacita sugar factory, whilst the recommendation domains for experimental farms continued to be used for many years, possibly still today.

The earliest indigenous soil surveyors at professional level included Harry Obihara and M. Nnodi. Obihara completed a substantial regional survey in Eastern Nigeria, jointly with British and Dutch expatriates, in 1964. Listed among nine senior officers working on the 1962 cocoa survey is J.U. Nwokoye. It is clear that professional Nigerian soils staff were taking up posts from the early to mid-1960s.

A second era of work followed after Nigeria's independence in 1960. The Land Resources Division, financed jointly from British aid and Nigerian sources, completed a series of resource inventories. Results were reported as *Land Resource Studies* and *Land Resource Reports*, covering large parts of Northern and Central Nigeria. These were at reconnaissance scales, 1:500 000 and 1:250 000. Each regional survey was a team effort, with publications under joint authorship. Among the leading staff were Michael Bawden, John Bennett, Anthony Blair-Rains (vegetation), Ian Hill, Dick Jenkin, George Murdoch, and Paul Tuley (agriculture). These were multi-disciplinary studies, covering climate, landforms, vegetation and soils, integrated as land systems, and translated into development terms through land evaluation and agricultural development possibilities.

Their nature is illustrated by the last and largest of these studies, *Central Nigeria*, covering 230 000 km^2, 25% of the country. It was surveyed over eight years by a team of twenty: three geomorphologists, six soil scientists, a soil erosion specialist, four ecologists, two rangeland specialists, a forester, two agriculturalists, and a consultant agricultural economist. The area is divided into six regions, e.g. the Bauchi Plains, the Jos Plateau. For each region the results are presented in three volumes: maps, agricultural development possibilities, and executive summary. The Western Savanna State region includes an innovation, a separate volume, *Atlas of Sample Blocks*. These are blocks of land 3.2 by 0.8 km, surveyed at 1:20 000 scale, showing the detail of local soil patterns which lie within the mapped units at smaller scale. The maps are of land systems, present land use, crop options, and agricultural development possibilities. These last are made up of eight classes, showing areas suited to integrated [i.e. smallholder] agriculture, mechanized farming, livestock production under specified management systems, and types of production and protection forestry. This is a comprehensive survey of land resources, converted through land evaluation into options for development. Viewing the outputs, one is left with the feeling that if government and aid agencies fail to proceed with agricultural developments, it will not be for lack of information.

In 1969-72 George Murdoch joined the team, joined by Nigerian collaborators Tayo Ojo-Atere, Emmanuel Odugbasna and Olu Olomo. They took sample areas of the land systems map and surveyed the component land facets at 1:125 000. This is one of very few examples of land systems being disaggregated into land facets.

The North East Nigeria project was distinctive in having a large collaborative effort, the partners being the Agricultural Research Station, Samaru. Staff from Samaru included two Dutchmen, Kees Klinkenberg in soils and P.N. de Leeuw in pastures and livestock. De Leeuw had acquired a remarkable insight into nomadic pastoral practices, combining a comprehensive knowledge of grassland ecology, livestock management, and the nature and problems of the Fulani people. He was later to take his unique expertise to the International Livestock Centre for Africa (ILCA).

It is difficult to know the extent to which use was made of these surveys. John Bennett writes:

> One of our main frustrations over the years was the fact that projects ended when the reports were presented, and we hardly ever had the opportunity to make follow-up visits. We therefore only had anecdotal information on the uses made, and this usually referred to third-party use as a source of natural resources information. If there was little that the local administrations could do with the reports, whether through lack of planning capability or budgetary support, there is strong evidence that the land resource studies provided valuable reference documents over the years, often for consultancy firms involved in competitive tendering. In 1998 I returned to Makurdi in Central Nigeria, now a city with two universities, and was introduced to the Dean of Forestry. He immediately linked my name with the series of LRD reports in his bookcase, and said they were still the only source of natural resources information for the area, which was very gratifying.

In the simplest terms, the existence of reconnaissance studies gave a flying start to proposed development initiatives, hastening them by the several years that would otherwise need to be spent on resource surveys.

Recollections

A standard question for surveyors to ask farmers, often through an inter-preter, is, 'What is your main problem in farming here?' Posing this in an isolated area on the Niger flood plain, the reply once conveyed back to Graham Higgins was, 'Elephants, sah.' He also reports: 'Typical Catering Rest House menu for visiting French surveyors from the adjoining Répub-lique de Niger: groundnut soup, stringy goat, yams and custard. Return visit by British: pâté, lobster, cheese board, profiteroles.'

Tony Smyth first met Higgins near the confluence of the Niger and Benue Rivers, on the site chosen for the Osara Experimental Farm:

> Typically, Graham was asked to do a survey only after the site had been chosen by government and most of the buildings constructed. On arrival I was quartered in the Agricultural Officer's house which lacked a roof, while Graham, his wife and a huge hound occupied the garage. The next morning we departed early to appraise the area and endure the hottest day of our lives. Before going far we asked a couple of labourers to dig a pit, left them to it, and wandered for five hours through the desolate landscape, making auger borings wherever rock or ironstone was not actually at the surface. The labourers were forgotten until we returned when, to our alarm, we could see only one of them, lowering a bucket on the end of a long rope. We approached gingerly and were horrified to see the digger at least 5 metres down a vertical pit in unconsolidated levee sand. Scarcely daring to move we whispered 'Come out, very carefully.' He cheerfully clambered out, more mystified than annoyed to be told to fill in the greater part of the pit. We imagine he is still dining out on the antics of mad Europeans.

When Bennett was in charge of the Central Nigeria Project, his field teams reported hostility from the local people:

> This developed into overtly threatening behaviour and we had to abandon activities. I was called to a meeting of the local chiefs, held in a rather tense and hostile atmosphere, and was strongly advised to leave the area p.d.q.. The local Federal Commissioner talked with the chiefs, then also advised me to make tracks. Apparently the cause was a sequence of events which had occurred at the outbreak of the Biafran war [a failed attempt at independence for the Eastern

Region]. Before the Biafran army broke out northwards they were preceded by reconnaissance parties, led by white men, posing as oil exploration teams. These had also dug holes and taken samples. I can't verify this but the local people seemed to believe it and we were left with a blank on our maps.

Harry Vine (1916-2004)

Harry Vine reached soil science in an indirect way. Educated as a chemist, in 1940 he was sent to Nigeria to look for alternatives to quinine for the treatment of malaria, and from there moved into soils. For the next eight years he worked as agricultural chemist. Then in 1948 he was seconded to Ghana for eighteen months and worked with C.F. Charter, learning about soil-landform relationships in the field. Returning to Nigeria he began the first survey of the country. By 1952 he had produced a provisional soil map of the country, writing in the preface that he 'adopted many ideas …with which he became acquainted during his work with Mr Charter.'

He also analysed results from an agricultural experiment maintained at Ibadan from 1922 to 1951. The Broadbalk Field at Rothamsted, UK, planted with winter wheat every year since 1843, is the classic example of a long-term field trial. In the tropics these are rare; they have to be funded from the core budgets of agricultural research stations, and there is an ever-present temptation to change to treatments or abandon the trial altogether in favour of perceived more immediate needs. The present-day emphasis on sustainable land use has lent a new importance to trials of this kind.

Vine subsequently took his knowledge and experience into teaching. He followed Hardy as Head of Soil Science at ICTA Trinidad, 1956-61, and then returned to Nigeria to the newly-established University of Ibadan 1961-65, being supervisor for its first PhD. During this time he wrote the soils chapter for Webster and Wilson's classic textbook, *Agriculture in the Tropics*.

Sierra Leone and Gambia

Small countries usually received a disproportionate amount of assistance in relation to their populations. The two smaller countries on the West African coast, Sierra Leone and the Gambia, saw some notable firsts and early studies of their kind, one of them a major development in survey methods.

In 1957-58 Martin Brunt spent the better part of two years in the swamps of both countries, making surveys directed at expansion of rice growing. At the time he was Land Use Officer to the Directorate of Colonial Surveys, a post that was to evolve into the Land Resources Division. The Gambia study was based on comparison of air photographs of 1946 and 1956, identifying changes in areas under rice and those suitable for further expansion. The work in Sierra Leone was a broader ecological study of the coastal swamps. The Colonial Pool of Soil Surveyors had been in existence for four years when Alan Stobbs joined it; his first secondment in 1958 was a semi-detailed survey (1:50 000) of the Boliland region of Sierra Leone. The accompanying text was produced in typewritten and staple-bound form, a hazardous practice which can lead to the work later being overlooked.

Subsequent work in Sierra Leone included a reconnaissance survey and evaluation of agricultural potential in 1963, led by C.J. Birchall, and an FAO/UNDP soil and land use survey of Eastern Province, completed in 1968. In 1974 American aid came in the form of surveys by the Agricultural Experiment Station of the University of Illinois (led by R.T. Odell), seeking a base for experience in Africa.

In the Gambia, it was observed that oil palm grew wild around the swamp margins, and in 1965-67 Ian Hill was sent to see whether plantations would be possible. It became clear that this was not the case, on environmental and economic grounds, and he was able to recommend that such development should not go ahead — in terms of money saved, one of the most valuable functions of soil survey. In fulfilment of the terms of reference, however, his report contained a map unique among resource surveys, showing 'Least unsuitable areas for oil palm.'

The major advance in methodology was one of the earliest examples of a new and welcome type of study, the comprehensive development feasibility survey. These are called feasibility studies in that the decision to invest has not yet been made, and are comprehensive in covering all aspects of land resource development. Four such studies were produced by the Land Resources Division at the peak of its activities, 1974-1976. Three were of areas in Nepal, Ethiopia and Belize, but one covered development prospects of an entire country. This was *The agricultural development of The Gambia: an agricultural, environmental and socioeconomic analysis* (1976).

These surveys demonstrated how survey and evaluation of the natural resource base could be extended through the whole sequence of development planning. The team involved, for overlapping periods, three soil scientists, a climatologist-hydrologist, two agronomists, two pasture/livestock specialists, a wildlife ecologist, a forester and a forest economist, and two 'socio-economists'. The survey operations comprised:

- Natural resource surveys: soils, ecology, climate, hydrology, and present land use.

- Technical studies: crop production (particularly groundnuts and irrigated rice), animal husbandry, and forestry.

- Social surveys: population, land tenure, social structure, and sample village studies.

- Economic analysis, covering market prospects and economic viability for the types of production (crop, livestock, forest) which the land appeared suitable in environmental terms.

In the report, nearly half the space is devoted to physical resources, a quarter each to technology and a quarter to economics. This reflects not the respective importance attached to these aspects but the nature of the data, in particular the great variability of physical environmental conditions, so often neglected in economics-based development planning. There are twenty pages of recommendations for development.

To this day these four comprehensive development feasibility surveys deserve wide circulation and attention. By successfully integrating surveys of physical resources with studies of the means for their development, they represented a new and distinctive contribution to methods of land resource survey, which could be more widely followed.

6

BIOGRAPHICAL INTERLUDE

THE REGIONAL ACCOUNTS can be interrupted by some biographies of giants of the soil survey world, who made major contributions to work in more than one part of the world. C.F. Charter began working on soils in 1931 and so would have qualified for inclusion in Chapter 2 as a pioneer. The most notable part of his career, however, was in the post-war period, so he represents a bridge between the pioneers and the golden age. Hugh Brammer served in the Colonial Service from 1951 to 1961, after which he had a long and distinguished career in what is now Bangladesh, first with FAO and then for the government of that country. His work thus extends from the period of Colonial surveys into an era of work by international agencies and locally-staffed survey departments. Jointly, their work extends over more than sixty years. Charles Wright divided a long post-war career between Central America and the Pacific islands. Very different from each other in personalities, they possessed in common enquiring minds, a dedication to the welfare of tropical countries, and a vast capacity for work.

Following these, short biographies are given of two British scientists, Anthony Smyth and Maurice Purnell, who began as soil surveyors and ended holding leading positions in, respectively, British and international institutions. The chapter ends with an account of some distinguished guests, European scientists who contributed to work in the Commonwealth and adopted British nationality.

Cecil Frederick Charter (1905-1956)

C.F. Charter — he was never addressed as Cecil but as 'Mr Charter' or familiarly 'Charter' — entered the world of soils indirectly. After taking the Natural Sciences Tripos at Cambridge he taught in grammar schools at

Tientsin, China, and then in Antigua. He must have demonstrated a capacity as a natural scientist for in 1931, at the age of twenty-six, he was asked by the government to undertake a study of the soils of Antigua and Barbuda, in the Leeward Islands. This reconnaissance survey was published in 1937, but in the meantime he had turned his attention to the mainstay of the economy, sugar cane, identifying specific soil requirements of the crop and collaborating in manurial experiments. Work in Antigua, Trinidad and Belize continued through the war years with eleven reports, 1940-1943, for the Trinidad Sugar-Cane Investigation Committee and the consultant firm of Scott Farnell. His twelve years of work in the West Indies is little known outside the region, but they supplied a self-training basis for his knowledge of soil-crop relationships.

Then came the event which was to take Charter to another continent and crop. In December 1944 he accepted an appointment as Soil Scientist to the West Africa Cocoa Research Institute. In 1945-46 he travelled extensively through the forest zones of Ghana, Nigeria and the Cameroons, studies which led to a report, *Cocoa soils: good and bad* (1948). In 1948 he was made head of the newly-created Soil Survey Division of the Department of Agriculture, becoming Director of the Soil and Land Use Survey Division in 1951.

Then followed the most productive period of his career. Regarded as a somewhat awkward customer by the administrative branch of government (and regarding 'admin' as a regrettable but necessary interruption to scientific work), he yet managed to get his own way. By 1954 he had set up a survey department of some twenty professional staff and 150 technical assistants.

Charter devised a method of soil survey for Ghana intended to make full use of local staff. Soils, vegetation and land use were recorded and sampled at regular intervals along traverse lines cut across the grain of the country, and soils were mapped by topographical associations. Local staff were trained to undertake each specialist task: setting out and measuring traverses, recording vegetation and land use, and sampling soils. The idea was to use large numbers of local staff to do the routine tasks, enabling the relatively expensive soil surveyor to cover more ground than he would otherwise have done. Within each topographical association, sample strips a quarter mile wide and one mile long (0.4 by 1.6 km) were selected and mapped in detail, to illustrate the complexity of the patterns occurring. This was nearly equivalent to the mapping of land systems and land facets.

With expatriate staff he gave no formal training, assuming they could do the job and treating them as equals. Such advice as he offered was given over a few beers in the evening. Hugh Brammer, Charter's colleague, said that 'the most remarkable thing about him was the way he trained the Africans, so that staff of limited education could do semi-professional work.' He was popular with both expatriate and local staff.

Three great characters in tropical soil survey:
C. F. Charter (left), Hugh Brammer, and a cocoa tree

Charter was keenly interested in the agricultural development of his adopted country, and devoted great effort to attempting to ensure that this would take place along sound lines. A starting point was mapping geographical regions, identifying correlations between environmental factors, and comparison with similar environments within and outside the country. The requirements of crops, actual and potential, should then be identified by setting up of trials. Arising from his experience in the West Indies he advocated establishment of a number of estates, which could employ improved methods, support experimental work, and so be of benefit to neighbouring smallholders.

In addition to the practical side, he maintained a keen interest in soil genesis and classification. Students of tropical soil science today will not realize how exciting the subject was in the 1950s and 1960s. Concepts of soil formation were being formulated on the basis of new field observations, and ideas exchanged internationally. Remarkably for someone directing a survey department, he continued to take an active part in these developments. Charter came to realize that some soils were not formed from the rock beneath them but from drift materials. He staunchly defended the genetic basis for soil classification. His ideas were based on the distinction between the acid oxysols (on which cocoa would not survive) and more fertile ochrosols. This terminology has not survived, but the concepts were incorporated into subsequent knowledge.

Charter was known for not suffering fools gladly, and his forthright manner of expression enlivens his publications. In 1950 'while spending my leave in Trinidad' the Colonial Office asked him to visit Guyana and report on the prospects for growing cocoa. The outcome of this study is a gem among consultancy reports: 'It may be categorically stated that … attempts to establish an economic cocoa industry in the Potaro District would prove, beyond any doubt whatever, a costly failure.' The soils were equally unsuited to suggested alternatives of coffee, cashew or dairying. His expressed views on the East African Groundnuts Scheme have been recorded above.

Further examples of his trenchant style come from two papers published posthumously in 1957. Writing on soil classification he blasts artificial classifications as being 'like the filing systems of Government Departments.' Most renowned is a pithy and concise article on the aims and objectives of tropical soil surveys:

> A number of soil survey organizations are being expected to discover areas, usually of more or less 'undamaged' forest, in which peasant production of such crops as cocoa, bananas, etc. will … be an unqualified success from the start. This is nothing less than turning soil scientists into pedological procurers who are given the task of finding attractive virgin lands for agricultural rape.

Given the conventions at that time on what could appear in print, for this to appear as a leading article in the highly respected international abstract journal *Soils and Fertilizers* was heady stuff.

Charter married twice whilst in the West Indies, but by the time he reached Ghana he was a widower. He was a chain smoker, which probably contributed to his death from cancer in 1956, during the full flowering of his abilities. After his death his deputy and successor, Hugh Brammer, collected a bundle of papers from his desk. These are an amazing legacy. There are 157 numbered pages, clearly handwritten in pencil, with abundant soil profile diagrams, landscape cross-sections and other illustrations. Three sections are headed, 'Classification of tropical soils — C.F. Charter', 'Forest and savannah soils', and 'Morphology of tropical soils (1952)'. Editorial marks in red have been added, such as 'Title page please' and 'Capitals', showing that these are intended as drafts for a book. The section on classification ends with, 'Card Index of Tropical Soils: author, date, title, types of soils, profile diagrams, photos, chemical data, countries,' anticipating modern databases. There is material here for a dedicated student to summarize for posterity.

His obituary in *The Times* ends, 'A prosperous agriculture in the country of his adoption is surely the memorial which he himself would have chosen.' Regrettably, this state of affairs was destined not to last. In 1957 Ghana became the second African Commonwealth territory to achieve independence, but over the years that followed, deficiencies of government were to lead to a decline in prosperity of its cocoa industry and agriculture in general.

Charles Wright (1915-1998)

Charles Wright was one of the great characters in the world of soil survey. The son of a glaciologist attached to Scott's Antarctic Expedition, he took a botany degree at Leeds and, gaining a First, was awarded a scholarship to visit academic institutions in Australia. This began with a passage on a P&O liner, with black-tie formality and the obligations which that entailed. By the time he reached Australia his bar bill had so decimated the scholarship that he abandoned plans for a research degree and, crossing over to New

Zealand, earned his living as an itinerant labourer on logging camps, construction sites and sheep farms.

At that time Norman Taylor was pioneering soil survey in New Zealand and took on Wright as a chainman and pit digger. Taylor recognized his keen eye for landscape, ecological relationships and soil management needs, and recruited him as a soil surveyor. Wright took part in the early years of the New Zealand Survey, including the siting of its headquarters at Lower Hutt.

In the Second World War he enlisted with the New Zealand Army Artillery as a surveyor, and served in the Solomon Islands. It may have been at this time that he adopted a New Zealand passport. He volunteered to dig trenches in the front line and took notes of soil profiles under enemy fire! His first paper was based on these observations. On return to work he was posted to, or more probably volunteered for, the Pacific island territories which the quirks of history had allocated to the country. His surveys of these included three small and exceedingly remote territories: Chatham (now Rekohua), Sunday (Raoul) and Niue Islands. Niue, in the Cook Island group, and Raoul, the principal island of the scattered Kermadec group, are respectively 15 and 10 km across.

Gaining promotion and fearing he might end up doing administration he landed a job as joint leader, with D.H. Romney, of a project in Belize (1952-54). Commissioned under the normal terms of reference of a soil survey, Wright and his colleagues spontaneously extended the scope into detailed consideration of land development and management. He next teamed up with Alun Jones of the Regional Research Centre for the West Indies to do a survey of Guyana.

Whilst proof-reading at the University of the West Indies, Wright received a cable peremptorily instructing him to return to New Zealand 'by the shortest possible route.' He opened an atlas and laid a ruler from Trinidad to New Zealand. He then followed his orders by taking a ship to Georgetown in Guyana. Alun Jones writes:

> He set out to walk the length of Guyana, starting on the Mazaruni River in the north and planning to go south by way of the Rio Grande, a tributary of the Amazon, to reach Manaus and thence up the Amazon and over the Andes to Ecuador. Moreover he planned

to walk the Guyana part barefooted to test what risks subsequent surveyors might run healthwise in the largely unexplored territory of the interior. He wrote to me later saying he had achieved most of his trek but had got behind schedule and had to fly one short section. All this happened long before the much-publicized Amazon expedition to find traces of Colonel Fawcett.

Whilst in Lima, seeking a trans-Pacific ship, Wright met a Chilean scientist who had been charged to set up a soil survey in his country. Without much persuasion he spent five unpaid months as soils adviser, house guest, and proselytising chef specializing in oatmeal porridge. He eventually got back to New Zealand twenty-two months later, and was immediately sent to Fiji on a New Zealand aid project, a national survey completed in 1958 and reported at formidable length.

Having started as a labourer he became Deputy Director of the New Zealand Survey. On leaving this he joined FAO, first on permanent staff but disliking the routine, changed to consultancy status and took up residence in Belize. Among projects in many parts of the world the most substantial was a survey of Western Samoa, comparable to that of Fiji in its thoroughness and, to the reader, somewhat excessive detail. I met him in Rome in 1978, when we were jointly engaged as consultants to determine the rest-period requirements (fallows, etc.) of world soil types for the FAO population carrying capacity study. Charles made estimates for Central and South America, I did the same for Africa and Asia, and we compared notes (for the same climatic and soil types, his were more cautious, i.e. longer recovery periods compared with cultivation).

Another posting was to Cuba. One day he was working with some Cuban surveyors at a road cutting when Fidel Castro chanced to pass by. The country's President stopped, enquired what they were doing, and decided to spend the night in camp with them. One of the Cubans expressed sentiments suggesting he was less than fully convinced of the benefits of Communism. Castro remained talking with this young man, arguing rationally and persuasively into the small hours of the morning to win over one single convert.

In Belize, Wright acquired a small farm on which he tried to put into practice the advice he had given over the years. His early work had given him a love of the country and a respect for its people, especially the Kekchi Amerindians. His air-conditioning system was to leave the house open on all four sides to let the breeze blow through, and his burglar protection system was to keep snakes.

In 1986 the British Overseas Development Agency undertook a survey of the Toledo District of Belize. The land resources team included Bruce King and Ian Baillie from the the UK, and Maria Holder and Juan Chen from the Belizean side. Baillie was told that:

> An old man, who had done a survey back in the 1950s, was to be their Soils Adviser. My heart sank, as I envisaged some dogmatic, tired Colonial. I could not have been more wrong. During the next five years, Charles and I worked together as part of a stable team. Even at the age of 75 he insisted on doing his share of pit digging.

Somewhat improbably, Wright was given an OBE, apparently for his early warning to the Belize government and British forces of Guatemalan military activity across the border. He was proud of the award but slightly sheepish about its name, Order of the British Empire. He once used this medal as security for a loan.

Wright's physical energy was amazing and he was totally oblivious of physical discomfort. He traipsed up and down hills, walking, talking, joking and smoking. He would get up at four o'clock in the morning and, with some rum for fuel, write the masses of clear prose which characterized his reports. His knowledge of soils, vegetation and agriculture was encyclopaedic, but he always insisted that farmers knew their land better than any itinerant soil surveyor. Capable of formality on occasion, he took a selective attitude to social conventions. Normally good-natured, he could occasionally lose patience with less energetic colleagues. His attitude to possessions and money was cavalier, seeming to regard a cash balance as a disposal challenge and an investment opportunity for friends in Toledo.

Leafing through back copies of the New Zealand *Soil Survey Newsletter*, Ian Baillie came across a reference to 'the legendary Charles Wright.' Ribbed about this, his reply was, 'You don't become a legend until you're dead. And I ain't dead yet.' He died at eighty-three, so now we can accord him his deserved status.

Hugh Brammer (b. 1925)

By today's standards, the Cambridge Geography Tripos in early post-war years was not technically very sophisticated, but it was wide-ranging, and produced graduates with enquiring minds and a dedication to field observation. Some of its staff, including Brammer's tutor, Benny Farmer, had gained experience in India and Sri Lanka during the war, and their teaching led to an awareness of how much was still to be learnt about the tropical world. Hugh Brammer (Bram) began with this background, his degree being interrupted by war service in the RAF.

When Charter founded the Soil and Land Use Survey of the Gold Coast in 1951 Brammer became his first staff member. It is possible that the older British style of recruitment, by word of mouth enquiry from Charter to the Colonial Office to the Cambridge Professor of Geography, Frank Debenham, played a part in his appointment. Brammer found himself in the best possible institutional situation, a small but growing department with an adequate operational budget and a Director who wanted his staff to get on with their job, untrammelled by administration, budgetary reviews and the like. He was also in the archetypal position during this era, of having at his feet a country where there was almost no systematic knowledge of the natural resources.

Brammer's first remit was the Accra Plains, the south-eastern corner of Ghana between the capital and the Volta Paver, important for the country's main export crop cocoa, and with prospects for irrigation. A detailed survey on a prospective irrigation area, Kpong, was incorporated. Over 2400 km of traverse lines were cut and recorded, and 500 profile pits examined. Staffing consisted of Brammer and in the early stages Charter himself, forty-eight semi-professional survey assistants, twenty drivers, messengers, etc., and a monthly average of 100 labourers. Preliminary investigations were undertaken in 1950 before Brammer's arrival but the main reconnaissance survey, 3900 km^2 at a scale of 1:125 000, commenced in April 1951 and was completed by February 1952. The bare statistics illustrate both the efficiency of Charter's survey methods and the energy of its leader.

Brammer became Director of the Soil and Land Use Survey following Charter's death in 1956, and remained there until 1961, overseeing the replacement of expatriates by Ghanaian staff. His *Soils of Ghana* (1959) remains as a framework to subsequent studies to this day.

He then joined FAO, which was beginning an immensely ambitious project, the reconnaissance survey of what was then Pakistan, West and East. From headquarters in Lahore, Brammer was sent to take charge of East Pakistan, 144 000 km², 'on my own at first.' Two initial tasks, of land and staffing, faced him. Disarmed at first by the flat nature of the terrain, he had to adapt the landform basis of soil survey to alluvial landforms. Then if the few expatriate surveyors were to make any substantial progress in so large a country, there was a need to train local staff, starting with twenty graduates in agriculture. Initially experiencing a certain amount of resistance he persisted, explaining what field survey was all about, and the purposes to which it would be put. 'Then one day I realized I had them in my hands.' Linked with the survey work was research, in conjunction with the academ-ically-minded Dutchman, Robert Brinkman, into the nature and formation of seasonally-flooded soils.

Work continued in East Pakistan until 1971 when civil disturbances, arising from the rejection by Bengalis of what they saw as domination by West Pakistan, led to the evacuation of the FAO team. Brammer was seconded to the World Bank in Washington to assist in preparing a Land and Water Sector study of Pakistan.

Brammer next served as Senior Soil Scientist in Zambia from 1972-1973, one of his duties being to instruct a Norwegian team in tropical soils and environments. Although in the country for only a short period he was by then in full flow of productivity, and produced a national map and three comprehensive, systematic accounts of the soils, one of which is, 'written for those who are not specialists in soils.'

He returned to the now independent Bangladesh in 1971, working with the Ministry of Agriculture and FAO as Land Use Adviser, Agricultural Devel-opment Adviser, and Project Manager for the FAO Agro-Ecological Zones study. This last involved extensive use of land evaluation, taking the results of soil survey and converting them into potential for crop production and agricultural development. In 1987 age necessitated retirement from FAO, but his position as an authority on development, recognized as such by the Bangladeshi government, led to frequent return visits, notably for the Flood

Action Plan (1989-1994). He devised small handbooks, written in terms that ordinary farmers and village leaders could follow.

All this time Brammer was writing. A capacity for clear expression is one of the strengths of the Cambridge tutorial system, but his copious output is unique. He wrote, like the novelist Anthony Trollope, in the early mornings. Between 1965 and 1977, jointly with Bangladeshi colleagues he produced thirty reconnaissance soil survey reports of the Districts, as well as numerous other publications and reports. During retirement he gathered these together and published them in no less than seven books, the last in 2004 at the age of seventy-nine. His objective in these was to place on record his vast knowledge of the country: 'The author is concerned that much of his knowledge and experience will die with him.' He has in mind his successors in the agricultural development of the country, primarily Bangladeshi scientists. There is also a wider value outside the country, in that these volumes present an unparalleled case study into how knowledge of land resources can be applied in agricultural development. Besides soils, the titles cover agricultural development possibilities, land use planning, research, flood protection, disaster management, and 'how to help small farmers'. Readers should not be put off by the mass of local detail; land resource scientists and developers will find these collected papers a mine of wisdom.

Brammer remained a lifelong bachelor, a colleague remarking that he was married to the soil. Talking to him, I asked for any unusual experiences in the field. The reply was that during the first reconnaissance trip with Charter, descending the ridge down onto the Accra Plains he saw in the distance what appeared to be an area of sand dunes, remarking on this to Charter who said nothing. 'But they turned out to be whale-like ridges of Archaean Sandstones.'

He also mentioned an occasion when he found himself in camp with 300 African staff, expected to organize ten survey teams; 'remarkably, it worked'. So it seems that he did not experience the more bizarre field experiences related by other surveyors. On asking for hobbies, the reply was much the same as that given by Cliff Ollier:

> Some tennis and badminton when younger, and a little gardening —
> until recently I looked after my mother's garden, until she died

recently aged 105. But what I found interesting was doing soil survey in three very different environments, and having to adapt. I was always a geographer, and fortunate enough to be able to spend my life as such.

Hugh Brammer was the most honoured of all mentioned in this book. He received an OBE from the British government on 'retirement' in 1987. Before that, in 1981 he received FAO's prestigious B. R. Sen Award, presented annually to the one field officer who had made the most outstanding contribution to the country to which they were assigned. But the recognition he doubtless values most had come in 1979, the Bangladesh President's Gold Medal for Agriculture. Among the community of natural resource scientists, Hugh Brammer became a legend in his lifetime.

Anthony Smyth (1927-2008)

After a Geology degree at Cambridge, Tony Smyth, as he was invariably called, joined the Colonial Service to carry out surveys in western Nigeria. Initially he and his wife Joyce lived at several bush stations where the climate was hot and humid, and there were few modern comforts. He subsequently became Director of Soil Survey at Moor Plantation, Ibadan. This apprenticeship — if you can call it that, it lasted ten years — taught him two things. The first was the essential geographical knowledge of soils in the field, and their relations with climate, geology and vegetation. This was published in one of the finest accounts, written jointly with Roy Mont-gomery, *The Soils and Land Use of Western Nigeria* (1962). Secondly, because the local economy was largely dependent on cocoa, he was responsible for developing one of the pioneer crop-specific studies, later to be expanded as as an FAO Bulletin, *Selection of Soils for Cocoa* (1966).

He joined the Soils Service of FAO in 1962, based in Rome but travelling widely in support of field projects. At that time, however, funding could still be found for Headquarters projects, employing consultants., and with Maurice Purnell he was responsible for organising the project which led to the *Framework for Land Evaluation*.[1] This became one of the most widely-

---

[1] I personally owed a debt to him for inviting me to the inaugural consultation on land evaluation and subsequently, to my pleasant surprise, being asked to co-author it.

used FAO manuals, and was followed by detailed guidelines on land evaluation for specific purposes.

With this record of field survey and organisation, Smyth was an obvious choice to take over in 1974 as Director of the Land Resources Division (LRD). He arrived during a period of expansion, when professional staff rose to over 60. Although soil survey still played a part, the work extended widely into ecology, forestry, and most importantly, economic analysis of proposed land developments. The results were set out in the 36 *Land Resource Studies*, unparalleled for their wide range and linkage between resource survey and opportunity for development. Smyth's later years at LRD were less happy, as government support for projects not externally funded was greatly reduced. He told me that in his bid for continued support he had quoted the words of a reviewer (me), that the output from LRD, 'constituted one of the greatest contributions to geographical knowledge produced by the UK', but with government attitude to research establishments at the time, this was in vain.

He retired in 1987 but remained active, serving on the Board of the International Board for Soil Research and Management (IBSRAM), and attending conferences. Tony was a genial man, combining wide knowledge of land resources, excellent staff relations, and a great sense of humour.

Maurice Purnell (1930-2015)

Following an Oxford Geography degree, in 1953 Maurice was recruited by C. F. Charter to the newly-established Soil and Land Use Survey of Ghana. This was one of the first British overseas territories to establish a strong national survey, and it is notable how many of their early staff went on to follow distinguished careers in soil science. This was followed by four years with Huntings, extending his experience to Iraq and Pakistan.

In 1961 he joined FAO, with which he was to spend the rest of his career. Initially this was with their field staff, providing soil surveys for projects in Brazil, Fiji, Myanmar (then Burma) and the vast surveys for the Jebel Marra area of Sudan.

The most distinctive was a survey, 1970-73, in Myanmar, where he worked with Russian colleagues, being able to communicate with one of them by means of a shared fluency in Spanish. In 2006, for the first edition of this book, in answer to my request for more information on this survey, Maurice wrote:

> Our project was the development, with irrigation, of the Sittang Valley — pretty huge area and, if I say it myself, I did some of my best field survey work ever, in difficult conditions, and trained the locals well too. We were constantly under armed guard to protect us from insurgents.

In 1977 he joined FAO Headquarters as a Technical Officer in the Soils Service of the Land and Water Development Division, with which he was to spend the rest of his career. FAO was nearing completion of the first *Soil Map of the World*, and the question arose where to go next to translate this into terms of development potential. A meeting was called at which the approach of land evaluation, developed by Netherlands scientists Klass-Jan Beek and Robert Brinkman, was adapted to FAO use, leading to the landmark publication, *A Framework for Land Evaluation* (1973). Maurice was to take a leading role in extending the Framework into detailed land evaluation guidelines for rainfed agriculture, irrigated agriculture, extensive gazing, and forestry. One could say that he became FAO's 'Mr Land Evaluation'.

Subsequently he turned his efforts to the more complex issues of land use planning. On one occasion he recognised its problems by reproducing M. C. Escher's drawing of men endlessly trudging up circular steps without ever reaching an objective.

On reaching FAO's retirement age in 1962, his services to them were by no means finished. He was active in the association of professional staff, concerned in particular with pensions. He was editor of their *Newsbrief*, bringing out an issue only a few weeks before his death.

Lacking personal ambition for top-level administrative posts, Maurice was especially helpful to younger staff, and hospitable to the many visitors to Rome. From the earliest field project he was also solicitous in training local staff. A former colleague wrote, 'He had the capacity to maintain a calm working climate in his team, and a sustained interest in their well-being.'

His leisure interests were art and wine. A lifelong bachelor, he was very proud of his family, following the lives of his brothers and sisters.

Distinguished guests

The Colonial Service gained from the services of a number of scientists born outside the Commonwealth who adopted British nationality. In three cases there were parallels in that their careers began as prisoners of war.

Stanislav Radwanski (Stan) (1923-1993) was born in Warsaw. He was sixteen at the time of the German occupation, joined the resistance movement, and became engaged in the most audacious and dangerous actions. Captured by the Germans three times, he escaped twice. The third time he was arrested, tortured and sentenced to death. The underground organization bribed the guards and instead of being executed he was sent to a concentration camp. From there he escaped by pretending to be a dead body, and was thrown out of a wagon on a pile of corpses. Recaptured in Austria, he was taken to another camp where he survived further severe punishment, which his two companions did not. Liberated near the end of the war, he joined the Polish army in Italy.

After the war he reached Palestine, where he completed his school education. In 1946 he arrived in Britain together with other Polish soldiers, as the Soviets were persecuting patriots and intelligentsia; news arrived that they were looking for Radwanski. He joined Guildford College to learn English and study chemistry, earned a place at Cambridge, and in 1950 obtained a degree in agriculture.

The following year he was spotted on one of Charter's recruiting trips and joined him in the fledgling Soil Survey of Ghana. Like many at that time, the most important part of his training was carried out in the field with Charter, whom his wife Estela describes as 'a demanding guru'. Her own wartime experience had been no less remarkable: sentenced by the Soviets to twenty-five years of forced labour in Siberia, then taken to the war front when the Russians needed Polish troops to fight the Germans, she later reached India where the British government offered hospitality to Polish soldiers and their families, met Stan in Cambridge, married and joined him in Ghana.

Stanislav Radwanski with the Chief of Sanhori, Ghana

The next ten years saw Radwanski take part in two of the leading national reconnaissance surveys in Africa, Ghana and Uganda. In Ghana, 1951-56, he took responsibility for the Upper Tano basin, north-west of Kumasi. Stan and Estela travelled together, staying at rest-houses. 'After close and unpleasant encounters, we developed an unfriendly and aggressive attitude towards snakes, tarantulas and mason flies.' Mason flies are unusual in being a soil indicator: they roll up pellets of mud to build their nests (sometimes in the corner of a room), and where they hover and drop, the soil is invariably wet.

When the soils staff dispersed after the independence of Ghana, Radwanski was transferred to Uganda, to join the survey team assembled by Chenery. On the way there, his ship was in the middle of the Suez Canal at the moment the Suez War broke out. The ship eventually returned to the Mediterranean and he took six weeks to reach Uganda via Gibraltar and the Cape of Good Hope. During this incident he had been in some danger because of his Polish nationality, so on reaching Uganda in 1956 he became a British citizen. He completed the survey and report for the area north of Lake Victoria, Buganda, and collaborated with Cliff Ollier on one of the classic accounts of a soil catena.

The Uganda project being successfully completed, he joined Huntings, where he was put in charge of soil and irrigation surveys of the arid zone, often saline, of the Lower Indus Valley.

By this time, invisible scars from his wartime experiences were beginning to affect his health, and he was able to take only short field assignments between spells of illness. These included semi-arid environments in the Jebel Marra, Sudan, and in Northern Nigeria, and findings he made in these regions were later to prove of much significance. In Sudan, Radwanski noted that soils under the apple-ring acacia tree (*Acacia albida*, now *Faidherbia albida*) were more fertile than in the surrounding land, and the crop yields higher. The same phenomenon was seen in Nigeria beneath an introduced Indian tree, the *neem* (*Azadirachta indica*). The local farmers knew about this, and deliberately planted their crops beneath these trees. The causes are nitrogen fixation, recycling of nutrients by leaf fall, and possibly animals sheltering under the trees. Radwanski's descriptions of the effects of trees on soils were published in 1967 and 1968, ten years before the invention of scientific agroforestry. He became a recognized expert on the neem tree, published further work, and kept in touch with research centres through correspondence.

Radwanski died of brain cancer in 1993. Seven tumours were found on the lesion of his brain resulting from his wartime treatment.

Jan Kowal (1920-1980), like Radwanski, was Polish by birth. He was nineteen when on 30 March 1940 he left home in occupied Poland and made his way to the Romanian border. He reached Romania by wading across a river, whence compatriots helped him to join the Polish army fighting in France. A diary of the time exists:

> I decided to leave because I felt it was my duty to fight for my country. My juvenile imagination viewed the war in terms of glory … Oh, how stupid! Not one of my dreams was fulfilled … The French have obviously lost the war … God has saved me miraculously! I met a Polish Jew who took me into his car and drove me to Bayonne.

From there he embarked on a cargo ship to Liverpool, where Polish forces were living under canvas. He requested transfer to the Air Force, and was accepted:

> I don't know exactly where we are [it was Kirkham, Lancashire] but we are very comfortable. We sleep in beds, there is hot water, and we eat four times a day from white crockery … The English are good comrades and their attitude to Poles is friendly, but it is difficult to communicate. At times I feel very lonely, but memories of my family and home help me to keep going.

Here the diary ends.

After this prelude, Kowal was accepted for a degree in agricultural chemistry at the University of Wales, Aberystwyth, 1943-1945. Such was the distinction of his later research that he was awarded three subsequent degrees, MSc, PhD and DSc. He began agricultural work with three years as managing director of a mushroom farm in Wales, marrying a Welsh bride, and becoming a British citizen in 1949. The next twenty-five years were spent in West Africa, first at the notable research station at Moor Plantation, Ibadan, then in the Faculty of Agriculture of the University College of Ghana, Legon, returning to Nigeria to the Institute of Agricultural Research at Samaru in the north of the country. It was in this last post that he collaborated with Amir Kassam in a synthesis of climate, vegetation and soils, *The Agricultural Ecology of Savanna* (1978). Although he was never engaged in mapping operations, his work was a source of inspiration to scientists working in West Africa.

'Give instruction to a wise man, and he will become yet wiser.'
Jan Kowal at a workshop, Nigeria

The idealism which had sustained Kowal during the war years was trans-ferred to his work, to which he was totally devoted. His daughter writes:

> My father harboured a great sorrow at being exiled from the
> country of his birth, but at the same time felt a great urge to move
> forward, and make a positive contribution to humanity — as he put
> it, 'To give back because I have received so much.' He was passion-
> ately idealistic, and totally devoted to his work. My parents did not
> enjoy the Colonial attitudes when they first arrived at Moor Planta-
> tion. My mother was expected to curtsey to the Director's wife and
> to wear white gloves.

During the civil war in Biafra, Kowal once hid an Ibo colleague in the boot of his car to help him escape a massacre, and he nursed his wife through a terminal illness. His papers show that at the time of his death he was working on another book, which would extend his synthesis of the environ-ment to the whole of Africa. The manuscript of the introduction begins:

None of the problems confronting mankind has greater implications than the current extra ordinarily rapid growth of the human population [and consequent] heavy demands on food supply and the resource base for human population.

Willem Cornelis Verboom (Bill) (1909-1989) was one of the most distinctive of the Dutchmen who served in the Colonial Service. Retaining a Dutch accent, he liked to remark, with a twinkle in his eye, 'I am a nationalized Englishman.' Verboom was born in 1909 in Bogor, Indonesia (then Buitenzorg, Dutch East Indies) where his father was Deputy Secretary of Agriculture, Industry and Trade. Sent to Holland for schooling, he return to the Indies to become a tea and coffee planter, by the outbreak of war owning three plantations. He spent his spare time hunting tiger and bear, and at one time owned forty hunting dogs.

Willem Verboom

On the Japanese invasion he did not surrender when ordered to but melted into the jungle with friends to start a guerrilla campaign. When an ultimatum was issued that ten Dutch prisoners would be executed for every day that they stayed at large the group surrendered. He was taken to Changi Jail in Singapore and subsequently to a camp in Japan, where 60% of his fellow prisoners died. Verboom was on his deathbed when the first atomic bomb was dropped on Hiroshima. Asked if they knew about this, he replied, 'No, but conditions improved immediately and the guards suddenly became very polite.'

After recuperating under American care he returned to the Indies to find his estates trashed and another war in progress, Dutch against Indonesian nationalists. He was wounded in action twice and received a knighthood (orange order) for bravery. The Dutch government capitulated and the Colonialists were left in the lurch.

Verboom jumped a ship to Australia where he worked in a jam factory and as a stockman, before taking a Diploma in Tropical Agriculture at Queens-land University. To have someone in the class with plantation management experience must have been a daunting experience for his teachers and fellow students.

He responded to a British government advertisement requesting the services of ex-servicemen in the African colonies. This led him to begin his professional career at the age of forty-four in Zambia, where he served as Conservation Officer 1953-61, adopting British nationality. Following Zambian independence the new Minister of Agriculture, Elija Mudenda, asked him to return to become Senior Conservation Ecologist 1967-74. During that time he led the ground-breaking Land Resources Division survey of the ecological resources of Western Province. In the interval between the two Zambian assignments Verboom took a contract with Huntings, as leader of the soil survey team on the Jengka Triangle project, Malaysia.

Bill Verboom was a tough, rugged man with a guttural voice and bushy beard. He would put this to effect when a crowd of women and children gathered round the camp. He would growl like a lion, the children would run away screaming and the women roar with laughter.

After retiring from his second spell in Zambia at sixty-five, he travelled the world to find a retirement spot, and chose a small property in Knysna, on the Garden Route in South Africa. He died in a car accident there at the age of eighty. When his son arrived to deal with the funeral he found that

Verboom had requested that his ashes be scattered into the ocean, to ensure that some of his molecules returned to Indonesia. A fellow ecologist and one-time Rhodesian politician, Allan Savory, used his political contacts to ensure a grand send-off with the ceremony conducted on a South African naval frigate.

7

CENTRAL AND SOUTHERN AFRICA

BY A QUIRK OF HISTORY, the British colonies in Africa which lay astride the equator, Uganda and Kenya, were called East Africa, whilst the term Central Africa was applied to countries 10-22 degrees latitude south, Malawi, Zambia and Zimbabwe. Cecil Rhodes and Sir Harry Johnston bear some responsibility for this situation. The countries of Central Africa temporarily became the Federation of Rhodesia and Nyasaland (1953-63), a measure unpopular with their African peoples, before moving to independence as Malawi, Zambia and (after an interlude as Rhodesia) Zimbabwe. Apart from delays caused by states of emergency, the Federal episode had little effect on staffing and activities in land resource survey. Because of the author's personal involvement, Malawi is covered in a separate chapter. The three 'High Commission Territories', Lesotho (Basutoland), Botswana (Bechuanaland) and Swaziland, formerly held a special political status, but again this had little significance with respect to land resources.

Zambia (Northern Rhodesia)

Why should Zambia, formerly Northern Rhodesia, have produced so many outstanding personnel, and been the site of no less than four original and distinctive innovations in land resource studies? It cannot have been the environment, which does not have the range and variety of, say, Kenya or Malawi. More probably it is due to their good fortune in recruiting staff of the highest ability, capable of thinking for themselves and given the chance, by their superiors, to do so. The first of these giants was the ecologist, Colin Trapnell, whose pioneering ecological surveys have been described in Chapter 2.

The second was William Allan, recruited as an entomologist in 1928 at the age of twenty-four. He went on field trips with Trapnell, and became interested in the question of how many people an area of land could support. The prevailing land use system at the time was *chitemene* or shifting cultivation. In a savanna environment the fallow period required to restore soil fertility after two to three years of cultivation is long, up to twenty years on the poorest soils. He expressed this as the land use factor, the ratio between years in the total cultivation-fallow cycle and years under cultivation, for example two years cultivation followed by eight years fallow gives a land use factor of $(2 + 8)/2 = 5$.

Allan perceived that by taking the per capita food requirement, the average crop yield on a given soil, and the land use factor, calculation would show how many people could be supported without decline in soil fertility. The observed population density of the Bemba people, for example, varied from 1.5 to 6 people per square kilometre, averaging 4.6. The land appeared to be in a steady state of long-term fertility. Once population increase had led to shortening of the fellow period below some critical level, fertility and thus crop yields would become lower, and a 'spiral of soil degeneration' would set in. Allan set out these views in a study of the Plateau Tonga in 1945. By then he had become Deputy Director of Agriculture, leaving in 1948 to become Director of Agriculture successively in Mauritius and Cyprus. His ideas were extended in *Studies in African Land Usage in Northern Rhodesia* (1949), and brought to wide notice in *The African Husbandman* (1965), an influential book which helped to show a generation of students the complexity of African agricultural practices.

Allan's concept of the land use factor was to prove fruitful. The German agricultural geographer Hans Ruthenberg modified it to the cultivation factor, R, the percentage of years under cultivation in the cultivation plus fallow cycle; thus two years cultivation and eight years fallow gives a cultivation factor $R=(2/ (2 + 8))*100 = 20\%$. (It is not clear if he intended R to stand for Ratio or Ruthenburg.) The concept of the cultivation factor needed to maintain soil fertility became the basis for an FAO study of population carrying capacity of developing countries. This capacity is, of course, heavily dependent on the level of external inputs, which in Zambia in Allan's day was zero.

The Plateau Tonga study was conducted jointly by Allan, Colin Trapnell, and Max Gluckman, anthropologist and director of the Rhodes-Livingstone Institute which sponsored the work. In a Foreword, Gluckman describes their method of work: they would rise early and do a long morning's work, then meet together in the afternoon to discuss findings. One day Allan remarked to Gluckman, 'I've been walking between you symbolically, because he [Trapnell] sees only the soil and the trees, and you see only the villages and people. Someone had to see both sides.' In the light of his detailed descriptions of peoples and farming methods, the comment on Trapnell is hardly fair.

Clear evidence of overpopulation led to the next innovative study. By the 1950s it had become apparent that the Ngoni people living on the Eastern Plateau close to the Malawi border could not support themselves on the reserve allotted to them. Crop yields were low, pastures in poor condition, and there was serious erosion including gullying, particularly in the *dambos* (poorly-drained grassed valley floors). Local officials were at a loss to impose soil conservation measures. The Ngoni Land Utilization Survey 1954-55 is a substantial survey covering soils, vegetation, agricultural systems, crop yields, livestock, and population trends. It leads to estimates of land carrying capacities, separately for each chiefdom, not only for human population densities but also cattle carrying capacities of the pasture land. There are ten pages of suggestions for rehabilitation, including resettlement. The most unusual feature of this little-known survey is that it was carried out not by research staff but by the District Officer, M.J.S.W. Priestley, and the Agricultural Officer, P. Greening. They must have drawn upon the accumulated knowledge of specialist staff and applied it, most skilfully, to the area they knew so well.

Moving forward in time past independence, the fourth distinctive work was an ecological survey of the Western Province completed in 1970 by the Land Resources Division, together with the Zambian government's Conservation Ecologist, Willem Verboom. Pasture and livestock specialists do not generally set great store on systematic surveys of pasture resources, and this is one of few examples. It begins with a standard description of land resources, which is followed by detailed material on grasslands, fodder utilization and cattle management. The work includes ecological survey maps at

1:50 000 for the Matebele and Mulonga Plains, with a very detailed tabular key listing grass, tree and shrub species.

Returning to standard soil survey, the country was no less fortunate in its staff. In 1954 Arthur Ballantyne was seconded from the Colonial Pool of Soil Surveyors, of which he was a founding member. His main work was a soil and land use survey of the copperbelt, where population increase drawn in by mining had led to a need for land use planning. Richard Webster became Soil Chemist, in charge of the soil survey, from 1957 to 1961. He introduced air-photograph interpretation, hitherto regarded there with some scepticism, as a means of speeding up survey. By that time, overpopulation was becoming apparent among the rural communities of the north-east, and surveys were needed of areas identified by the administration as potentially suitable for resettlement from the overcrowded reserves. Webster also joined a pasture ecologist, William Astle, in a land classification of the Luangwa Valley, one of the best and most accessible land system surveys. Webster's subsequent career encompassed terrain evaluation for military purposes, statistical studies of soil variability, and Editor of the *European Journal of Soil Science*.

The country was next to benefit from the knowledge, dedication, and capacity for work of Hugh Brammer, seconded by FAO on a two-year appointment 1972-73. His mandate was to advise a newly-arrived Norwegian team on tropical soils, but he took the opportunity to compile comprehensive accounts of the soil resources of the country. Throughout the post-war period, detailed soil surveys were being conducted: thirty were published in the period 1967-75. Some titles indicate their range of purposes: surveys of regional research stations, a tobacco scheme, an extension to a sugar estate, a proposed refugee settlement area, and an irrigability classification.

Twelve of these local surveys were carried out by Barry Dalal-Clayton during two spells in the country, the first as Land Use Planning Officer 1972-75. Finding himself initially as a two-man team under Hugh Brammer, he would not have been allowed to spend much time in the office. They went together into the field: 'Effectively I was given a crash degree course from a highly-experienced tutor.' Clayton's major contribution to soil science was an analysis of the sandveldt soils, which are so widespread, difficult to manage, and poorly recognized on international classification systems. He was also asked to do some land capability surveys, a system adapted from the United States and widely used in the Central African territories. 'I had no training in this having done botany at university, but looked through the Zambian land use planning manual and set off. I was greatly helped by the Zambian Agricultural Assistants who had done it for years under various expatriates.' Clayton was later to become Director of Strategies for the UK International Institute for Environment and Development (IIED).

Recollections

This account may be ended with a personal recollection which illustrates Clayton's comment about the value of local help. In the course of a short visit as external examiner to the University of Zambia, I was invited into the field by a soil surveyor allocated the extreme west of the country, an area underlain by Kalahari Sands. Almost all the soils consisted entirely of sand, a fact which may have affected his psychological health. On one occasion he sat in a Land Rover, stared at an air photograph for some moments, then muttered an expletive and lit another cigarette. I was aware that, relative to the orientation of the vehicle, the photograph was upside-down. The driver leaned across, perceived this fact, appraised our location, let out the clutch and set off in a beeline across the bush, arriving at the intended destination. This driver had also gathered that it was desirable to find anything other than sand. The village lay beside one of the circular depressions which characterize that part of the country. Without instruction he rushed off, hired a dugout canoe, and used the soil auger to punt across the water. Plunging the auger over the side, he held up the contents into the air, shouting, 'Clay, Bwana!'

Zimbabwe (Southern Rhodesia, Rhodesia)

The context for agriculture and land resource studies in what is now Zimbabwe was distinct from that of other countries. Separated from Northern Rhodesia in 1911, Southern Rhodesia became a self-governing colony in 1923. In the terminology of the time it was a 'white settler' country, with African farming restricted to Native Reserves and Native Purchase Areas. From 1953-63 it became part of an ill-fated political experiment, the Federation of Rhodesia and Nyasaland. When in 1964 its partners achieved independence as Zambia and Malawi, the British government refused to accept the entrenched position of the white minority, a state of affairs followed by a Unilateral Declaration of Independence (UDI) as Rhodesia in 1965. International recognition as Zimbabwe followed in 1980. Following unacceptable actions by its President, it was suspended from the Commonwealth in 2003.

The politics of these changes need not concern us, except insofar as they are relevant to land resources. Staff lists are in the normal Colonial form, although some staff were of South African origin. The Veterinary Department was larger than the Agriculture Department through the inter-war period. During the period of Federation there was a limited amount of cooperation between the three component countries, although this did not extend to joint agricultural research institutions and services on the pattern of East Africa. One outcome was a *Federal Atlas* at 1:2 500 000 scale, which served a function in linking the mapping units for the three territories. During UDI, Rhodesia became somewhat isolated, as scientists from other countries were discouraged from making visits.

The first national soil map, at 1:1 000 000, appeared in 1955, produced by R.G. Thomas and B.S. Ellis. Its sources are given as the Geological Map of 1946, together with local soil surveys and reconnaissance. Ellis had long experience of the country, having been appointed as Chemist in 1928. There are thirty mapping units, described in terms of rock type, vegetation, climate, and soil properties, for example (slightly abbreviated):

> Soils on basic igneous rocks; under deciduous open mixed bushland; medium-to-low rainfall and hot temperatures; dark reddish brown clays, slightly acid, over similar clays with numerous ironmanganese concretions.

Comments have occasionally been made that the soil map looks surprisingly like the geological map coloured differently. This may be unduly cynical. Ellis wrote an article stating that in Africa, over a wide range of humid conditions, parent material dominates soil properties and is therefore 'the most practical method of classification.'

The two most distinctive contributions to land resource survey, however, are in areas other than soil mapping. *An Agricultural Survey of Southern Rhodesia* (1960) is ground-breaking in its integration of natural resource information with the economics of farming systems. It is in two parts, initiated and coordinated by the Director of Research and Specialist Services, R.R.

Staples. Following discussions in 1950, a team of three was set up, consisting of an agricultural ecologist (originally J.M. Rattray, later replaced by Vane Vincent), a pedologist (R.G. Thomas) and an agricultural economist (R. Anderson). Part 1, *Agro-ecological survey*, is focused on a map at 1:1 000 000 showing 'Natural regions and areas' of which there are ten. Although delineated partly on the basis of climate and soils, these are defined in terms of farming systems, e.g. 'IIA Intensive grain crop production supported by livestock'. Two other features are distinctive. The first is agroclimatological, consisting of maps which show 'planting rains', the dates on which reliable rains can be expected with 20, 40, 60 and 80% probabilities. The second is a map of estimated potential arable land, the percentage of total land which can be cultivated (e.g. Class 1: >70%; Class 2: 50-70%; ... Class 6: 0-10%). I know of no other country where a survey-based estimate of this kind has been attempted.

It is Part 2, *Agro-economic survey*, which is the most innovative. Taking each of the agro-ecological areas as a basis, farming systems on them are reviewed. For each there are tables of average cropping mixture, and for each crop, the planted area, yield, value and sales are given. This leads to a map of Economic Regions and Areas, e.g. '2C Tobacco farming, 5A Cattle ranching'. A qualification is that this uniform treatment is applied to the European farming areas only. African agriculture is covered in a separate chapter, with emphasis on problems, especially erosion.

Clearly, this is a two-stage survey rather than being integrated. It is of interest in demonstrating the potential for linking natural resource information with economics. Owing to the isolation which followed UDI, it is regrettably little known.

The second distinctive contribution extends beyond survey into the area of soil conservation. This is the work of the Department of Conservation and Extension, CONEX (later becoming AGRITEX). In the post-war period Zimbabwe became the leading country in Africa for soil conservation. The basis was a system of contour bunds, earth banks and ditches running along the contour to intercept runoff, with grassed waterways to carry the water down to the streams. It was applied both on estate land and in African reserves. Flying over the countryside in the 1960s an amazing sight met the eye, a network of bunds and waterways looking for all the world like a diagram from a conservation textbook. In the 1970s CONEX developed the Soil Loss Estimation Model for Southern Africa (SLEMSA), an adaptation of the well-known US model more suited to subtropical and tropical African conditions. Those responsible were Henry Elwell, an agricultural engineer from Southampton University, and Michael Stocking, originally an Oxford University geographer. They drew upon data from the largest network of erosion measurement plots outside the United States, set up by Norman Hudson.

Linked to this work on conservation was a method of Land Capability Classification, also adapted from the standard US system, which assigned land to different permitted uses: arable, limited arable, grazing, forest, conservation, etc. This was widely applied in field-scale survey for opening up new land. It was later adapted, incorporated into land use planning handbooks, and applied in both the other countries of the Federation.

The leading role in the conservation work of CONEX was taken by Vane Vincent, a South African by origin. An important contribution was made by the extensive research into soil conservation, and advocacy of its application by Norman Hudson, working at what was then the University College of Salisbury (Harare) (see Chapter 13). An earlier source of ideas and methods was a postgraduate course on soil conservation given by Professor John Phillips at the University of the Witwatersrand, South Africa. Phillips was a leading figure in applying the ecological approach to land use planning and management, and his book *Agriculture and Ecology in Africa* (1959) is of lasting value.

Apart from Vincent, personnel in soil survey and conservation during the Colonial and Federation periods were of British origin. The country was later than most other African states in appointing indigenous Zimbabweans. The first was Kingston Nyamaphene, who after a spell as Soil Surveyor 1978-81 led a distinguished career in academic administration.

Botswana (Bechuanaland), Lesotho (Basutoland), Swaziland

Three countries of Southern Africa have in common that they were once administered by provinces of South Africa. They were taken under British 'protection' between 1869 and 1903, as Bechuanaland, Basutoland and Swaziland becoming known as the British High Commission Territories. Independence was achieved between 1960 and 1968, Bechuanaland becoming Botswana, and Basutoland (six years after independence) Lesotho.

The history of these three territories is not irrelevant to the development of their natural resources. For reasons both economic and political, they have been particularly favoured as destinations for British aid. Economically, whilst they have some mineral revenues, they lack export revenue from estate agriculture. Politically they are adjacent to South Africa, which in the post-war years was under the apartheid regime. The former remittances from migrant labour in South African mines diminished over the period under review. Hence there were many development projects, with the surveys needed for their efficient and sustainable application.

Botswana (Bechuanaland)

Botswana has been the site of relatively numerous studies, a fact not unrelated to its stable government. The Land Resources Division conducted a survey of the *Land Resources of Eastern Bechuanaland* in 1963, one of the few to be completed when they were still called the Forestry and Land Use Section of the Directorate of Overseas Surveys. Surveys of soils, land use prospects and irrigation potential followed, with Michael Bawden, Ian Langdale-Brown and Anthony Mitchell the leading figures. In 1977 David Eldridge was initially appointed as a veterinary officer, but when soils work was called for, he made use of his training in geography and started a soil survey team, joined by FAO staff and local graduates. In the semi-arid country, the government was keen to identify pockets of potentially arable land, the Arable Lands Development Project, for which farm sites were surveyed. Later than in most countries, some small-scale mapping based on satellite imagery was also conducted.

In the 1980s FAO carried out a series of projects in land use planning for sustainable agricultural development. This was focused on methods of natural resource assessment and land use planning appropriate to smallholder farmers in a semi-arid environment, and included a training element for twenty local graduates. Among the outputs was a national soil map at 1:1 000 000 scale and, based on a crop yield simulation model, national maps showing suitability for various crops under specified management conditions, an outstanding example of land evaluation.

As a semi-arid country, livestock production is a major element in the economy. The pastures were utilized both by ranches and tribal grazing peoples. The government commissioned a survey of how the grazing lands should be divided between these groups of users, the Tribal Grazing Lands Project. For this purpose, a nationwide consultative exercise was conducted, based on 'Radio Learning Groups'. Regrettably the report on this, *Lefatshe la Rona: Our Land*, is notable for the total absence of mention of any kind of land resource survey.

Lesotho (Basutoland)

Complete coverage of Lesotho at reconnaissance scale was achieved by the Land Resources Division, led by Douglas Carroll, C.L. Bascomb and Michael Bawden, in the 1960s, presenting the results as a broad land resource study together with a technical bulletin on soils. The mountainous and deeply dissected parts of this country must have provided some spectacular views under stereoscopic air-photograph interpretation. A six-volume appraisal of the irrigation potential of the country was carried out by Binnie & Partners in 1972.

Swaziland

Swaziland was for many years the province of one man, George Murdoch, who arrived there in 1955 and remained for fifteen years. His major work was *Soils and Land Capability in Swaziland* (1968) with a map at 1:125 000, twice the usual scale for reconnaissance surveys. Whilst assisted for three-year periods by Jacobus Andriesse (Netherlands) (1958-60) and Ian Baillie (1965-68), Murdoch essentially, in his own words, 'surveyed the whole country.' In 2001 he wrote, 'My outdated maps are, to my chagrin, still in use; can't somebody revise them, please?' which could be taken as a tribute to the usefulness and reliability of his work. He later joined with the team from the Military Engineering Experimental Establishment (MEXE) to produce an *Atlas of Land Systems in Swaziland*. In the late 1950s, what in other territories would have been the Department of Agriculture was termed the Land Utilization Department, headed by a Director of Land Utilization, C.J. Van Heerden (South Africa).

Irrigation was an active concern in all three of the former 'High Commission Territories'. The Colonial Development Corporation (CDC) sponsored an early irrigation survey in Swaziland (1961) by Huntings. There were a number of later surveys of water resources development by consultant companies, including in Swaziland by the Engineering and Power Development Consultants (1970). A model for more detailed appraisal of irrigation potential is Murdoch and Andriesse's 1964 survey of the soils and irrigability of the Lower Usutu Basin. Surveys for river basin development continued after independence, notably a study of irrigation potential in the Komati Basin, carried out by the classic combination of a natural resources consultancy, Huntings, and an engineering contractor, Alexander Gibb.

In developing countries as a whole, mapping of grasslands and their management are very much less common than soil surveys. An exception is a 1967 study by J.H. I'ons, *Veld Types of Swaziland*, with an accompanying guide on pasture management for extension workers and farmers.

More recently, Swaziland became active in land evaluation, producing crop-specific studies of suitability for cotton, pineapples and sugar, and an evaluation for plantation forestry.

George Murdoch (1930-2005)

George Murdoch was among those who founded a lifetime career in soil survey on the basis of experience in one country, Swaziland. At George Watson's School, Edinburgh he was not fond of games, but laid the foundation of a formidable memory for names by compiling lists of cricket and football players. National Service was spent in the RAF, where he became a pilot. His education was ideal for soil survey, combining the best of two worlds: a broad knowledge of natural resources from a geography degree at Edinburgh, and specialist study of soils at postgraduate level.

He became a founding member of the Colonial Pool of Soil Surveyors, posted to Swaziland in 1955. Unlike other members of the Pool, who moved through several countries, he remained for fifteen years. His wife, Eunice, was a cartographer, so they served as a team. He came to know every inch of the country, not only its landscapes and soils but its history, and could recount the ancestry of many of the chiefs. He acquired a great affection for the people, and took pains to explain his work in terms which farmers and development planners could understand. He was as happy to help individual farmers as to collaborate with international missions.

Midway through his service there he was awarded the MBE. Many years later a former Prime Minister of Swaziland learnt that he was on a visit to the country and asked to see him; affairs of state were put on one side as they discussed Murdoch's work on soils which the Minister had consulted as a student.

On return to the UK in 1969 Murdoch joined the Land Resources Division, and in 1975 became Head of the Soil Science and Geomorphology section. When he was working on a soil survey in Western Nigeria, the consultant company, Booker Agriculture, was developing a sugar estate on the banks of the Niger. Murdoch and some colleagues appeared on a boat round a bend in the river and introductions followed. This led to him joining Booker as Head of Land Resource Services from 1979-93.

At Booker, his knowledge of where to find specialists on any subject became legendary, and when a proposal was made to start a database of consultants, someone remarked, 'Why bother, we've got George.' He became a principal co-author of the *Booker Tropical Soils Manual* (1984), widely used both within and outside the company. He developed crop-specific land suitability systems, incorporated in the manual and subsequently one of the foundations of the FAO crop requirements database.

I had an example of his remarkable memory when writing in 2001 apropos of this book. In reply to my enquiry he sent the names and nationalities of 120 soil surveyors who had worked in British territories. These were neatly handwritten from A to Z, without corrections; I fully believe he held this list in his head in alphabetical order. He wrote, at the age of seventy-one, he was 'just off to Egypt on a mission to expand arable land using latest techniques of drip irrigation.'

Swaziland remained his principal interest and in 1991, jointly with the High Commissioner in London, he founded the Swaziland Society. He was a dapper dresser, sometimes arriving for work in London in a powder-blue safari suit. With a soft Scottish voice and a keen sense of humour, he was courteous, open and friendly, and many colleagues benefited from his advice and encouragement. His first wife died in 1991 and three years later in Kenya he married Muthoni, who was a source of great happiness in his retirement.

8

MALAWI (NYASALAND)

MALAWI, BEFORE INDEPENDENCE NYASALAND, is given a separate chapter because of my personal involvement over forty-five years. It can be considered as a case study, showing the development of land resource studies in somewhat more detail than space has permitted for other countries.

It is among the smaller African countries, half the size of the UK, but has more varied landscapes and resources than its larger neighbours. Much of its area is formed by the 'African surface', fragments of the gently undulating plateau at 1000-1500 m altitude which covers vast areas in Central and Southern Africa. Rising above this surface are six flat-topped, steep-sided massifs rising to 2000 m. Dropping below it are the dissected escarpment zones of the African Rift Valley which traverses the whole country. In the north and centre, the floor of the Rift Valley is occupied by Lake Malawi (formerly Lake Nyasa, which meant 'Lake Lake'). In the south, it is formed by the alluvial plain of the River Shire. Standing at Malawi's lowest point, 70 m, in the Lower Shire Valley, it is sometimes possible to see its highest, the summit of Mount Mlanje at 3002 m.

The people who settled the country found fertile areas which rewarded cultivation. For geological reasons the Lilongwe Plateau in the centre and the Zomba Plateau in the south have more fertile soils than most of the African surface, and because of the varied relief, rainfall is relatively high and reliable. As a consequence, the population density was, and has remained, three to four times higher than in any of the neighbouring countries. Paradoxically it is the richness of its natural resources which has led, through population increase, to its agricultural problems: small farms, soil degradation, and low crop yields. It is now among the countries recognized by the UN as 'least developed', that is, the world's poorest.

Nyasaland became a British Protectorate in 1891 (as British Central Africa) and a Colony in 1907. Then in 1963 came a political experiment, when it was joined with Zambia and Zimbabwe to form the Federation of Rhodesia and Nyasaland. However, suspicion of dominance by Zimbabwe made it unpopular with the people from the start. This led to political agitation and a State of Emergency in 1959, to be followed by an intermediate stage of self-government in 1962, and independence, as Malawi, in 1964. The period of Federation was undoubtedly a failure from a political point of view, although it led to some useful scientific cooperation between the three territories.

Soil chemistry

Soil survey in Malawi had gained a flying start through the pioneering efforts of Arthur Hornby (Chapter 2). Activities resumed after the Second World War with the appointment of Vernon Cutting as Analytical Chemist in 1949, who established a laboratory at Bvumbwe, near the commercial capital, Blantyre. The basic work of this was soil analysis on which to base fertilizer recommendations, both for the country's small estate sector (tea, tobacco) and for the African smallholder farming which occupied the greater part of the area. Later, when a national reconnaissance survey was started in 1959, the Bvumbwe laboratory was able to supply it with analytical services. The earliest Malawian professional soil scientist, at a time when local graduates were very few, was Walter Chona, appointed Assistant Chemist in 1959.

The Lower Shire Valley project

In the 1950s there was a major survey for a proposed drainage and irrigation project. This was in the 'lower river', or Lower Shire Valley, an alluvial area at 100-200 m altitude. It is exceedingly hot, has the highest density of mosquitoes to be found anywhere in the tropics, but has some rich black soils (vertisols) which support a good crop of cotton. The circumstance favourable to irrigation is that Lake Malawi, the bottom of which drops below sea level, forms a large natural reservoir, comparable to that of the (much shallower) Lake Victoria for the White Nile. There would be no need for the expense of constructing a storage dam, only a small control barrage. The land which appeared suitable for irrigation was Elephant Marsh, which would first require drainage.

Sponsored by the Crown Agents, the engineering company which reported on the potential for utilizing the River Shire was Sir William Halcrow and Partners. Their main report covered hydrology, flood control, and drainage, not forgetting entomology; on the economic side there was the standard material of a pre-investment study, estimated costs and incomes. For soil survey and land classification, Halcrow subcontracted to the leading company in this area, Huntings. The Elephant Marsh survey was a classic example of the collaboration between consultants in engineering and agriculture needed for irrigation and drainage studies. It is, however, fortunate from the viewpoint of wildlife conservation that this project was never implemented.

Ecological studies

Linking early work with the modern period is a delightful book by Jessie Williamson, *Useful Plants of Nyasaland*. Although published in 1955, the material had been collected in nutritional surveys 1938-43. She continued to collect material until an expanded second edition, as *Useful Plants of Malawi*, appeared in 1974. Illustrated by her own drawings, this ranks among the best works of its kind. For half a century it has been a treasured source of information for newcomers to the country.

Survey activities by the Department of Agriculture resumed after the war with the appointment of George Jackson as Ecologist in 1950. For the next nine years he found himself largely on his own, scientifically, with the country at his feet. Ecologists have to begin by identifying the plant species present and forming a herbarium collection. This leads on to the objective, identifying the vegetation communities and their composition. Jackson made the first synthesis as a *Preliminary Ecological Survey of Nyasaland* (1954). This is comprehensive and concise, although its usefulness and impact was reduced by extreme inaccessibility, being published in the proceedings of an Inter-African Soils Conference. Four years later, drawing upon outside help for identifications, he produced *An Annotated Check List of Nyasaland Grasses*, and at the end of his period of service, an account of climatic regions. Finally, lacking a soils colleague Jackson turned his hand to soil survey, developing a simple but practical technique for detailed studies. These also are little known, appearing in the Annual Reports of the Department of Agriculture. His final contribution, in the 1950 Report, is headed:

DIVISION OF ECOLOGIST

Ecologist in Charge: G. Jackson

… which is certainly true.

After the return to the country from self-imposed exile, in July 1958, of Dr Hastings Kamuzu Banda, later to become Life President, anti-Federal agita-tion steadily grew. Jackson disliked the atmosphere of mistrust between government and people which followed this, in such contrast to the friendli-ness of the people which he had earlier experienced. He was particularly hurt by the abolition of the local Boy Scout movement, to which he was committed, but which was incompatible with the politically-based youth organization, the Young Pioneers. His last duty cannot have been amenable. Because of his language ability, during the 1959 State of Emergency he was asked to read government news bulletins on the local radio, since no African could be trusted to do this without breaking into propaganda for the Malawi Congress Party. The news went out in fluent Chi-Nyanja, but with a Shef-field accent.

Jackson retired from Malawi, disillusioned, the same year, moving to a lectureship in Nigeria. Among the more unusual research projects he under-took there was tracing the movements of nomadic peoples by identifying seeds from the droppings of their cattle, a survey his colleagues referred to by an unquotable title.

The Agro-Ecological Survey

By the mid-1950s it was becoming apparent that Malawi lacked a reconnais-sance soil survey, comparable to those in progress in other colonies. The Annual Estimates of Expenditure for 1957-58 contain the item, 'Soil Surveyor, new post for Land Use Team, £937.10s to £1850, offset by reduc-tion of one Agricultural Chemist'. What part was played in the decision to make this appointment by Jackson himself, the Director of Agriculture,

R.W. Kettlewell, or Herbert Greene as Natural Resources Adviser to the Colonial Office, is not known; Robert Green, a Conservation Officer at the time, believes the initiative and foresight came from Kettlewell.

The manner in which the appointment was made is of interest in showing how attitudes to duration of service were very different from those of today. In 1957 a Cambridge geography graduate, about to complete a PhD in geomorphology, was interviewed and offered the post, but replied that he would accept it but only on three-year contract terms. Jackson told his director that this was unacceptable, as a natural scientist would not be of much use for his first three years in the country. The post lay vacant, until a year later the same candidate made an enquiry to the Colonial Office, and this time accepted the post on permanent and pensionable terms. Soon afterwards, from the 1960s onwards, governments and institutions would not dream of offering other than a fixed-term contract.

The next part of this account becomes a personal memoir, as the keen young man appointed was Anthony Young.

The normal passage for Colonial officers was still by sea, and I sailed from London in the *Braemar Castle* on 19 November 1958, having with difficulty persuaded the Colonial Office that my wife might accompany me. I spent time on the boat learning Chi-Nyanja, since officers could not receive an annual increment before passing a language examination. After a month on the ship I reached Blantyre by train from Beira on the coast of Mozambique, and thence to the Department of Agriculture Headquarters in the capital as it then was, Zomba.

A short but valuable handover from George Jackson took place against a background of civil protest. We went on a joint visit to Kachere, a village in Dedza District, Central Province, which had been earmarked for a land reorganization scheme to reduce the excessive fragmentation of holdings. In the field together, Jackson demonstrated his survey method: traverses by bucket auger, laying out the successive cores along the ground. To set soil boundaries he also made much use of indicator plants, trees or grasses which indicated wet or shallow soils (except, or so it seemed to a non-ecolo-gist, where they were found growing on dry or deep ones). He also became quite clear, by talking to young men in the village, that we were engaged on a project which would never be implemented owing to anti-government feeling.

On 3 March 1959 a State of Emergency was declared. Government officers whose services could be spared were seconded to the police, to stand in for routine duties whilst the regular staff were engaged in security operations. Among those involuntarily seconded was the Soil Surveyor. Compared with other African countries, Malawi's State of Emergency was short. Moves towards the break-up of Federation, elections, self-government and independence were soon under way, and normal government activities, including agricultural research, resumed by June 1959, an interruption of some three months.

Soil Survey of Malawi: the complete team 1958-1962, in front of the former Headquarters of the Department of Agriculture. From left, Khanyera (driver), Anthony Young, George Mandala, Reginald Mwangala (draftsman)), Evans Gunda

In 1959 I undertook two further local surveys, which can be seen in retrospect as training exercises in highly contrasted environments. The first was for a proposed smallholder tea scheme in a forest area on the shore of Lake Malawi near Nkhata Bay. There was an existing, very isolated, European tea estate, and the idea was that its processing facilities would be shared by adjacent African smallholders. The second survey was in the Lower Shire Valley.

Looking back on the map and report, one might comment that to distinguish nineteen mapping units in this small area showed excessive enthusiasm. The following year a *Preliminary Soil Map of Nyasaland*, scale 1:2 000 000, was produced, based on Hornby's pre-war maps of the Central region, a vegetation map by Jackson, and my early travels.

In July 1959 a start was made on the project for which the Soil Surveyor had been recruited, the Agro-Ecological Survey of Malawi. On being given the go-ahead I announced that it would begin with Northern Province and work its , way southwards, and in July set off for the most remote northerly part of the country, Chitipa (formerly Fort Hill).

For the next two and a half years a routine was followed, basically as set out in Chapter 3 above, for successive Districts of the Northern and Central Provinces. The base maps employed were at 1:250 000, and air-photograph coverage (black-and-white) at 1:25 000 or 1:40 000. Some were from 1947-1948 RAF air photography, flown in rings at constant distances from a central beacon, giving print lay-downs which were circular. Stereoscopic air-photograph interpretation was carried out in the standard way, in the first instance drawing boundaries between all regions of differing appearance, even if one did not know the reason.

"Listening to the soil": strength of reaction after
adding acid indicates degree of calcareousness

The next step was somewhat innovative at the time: to write to the agricultural officer of the District, asking for soil pits to be dug at sites of fertilizer or variety trials, or other sites where agricultural performance data were available. These became key sites for field descriptions of soil profiles, and type sites for soil series — nowadays they would be called benchmark sites. Apart from these, other field observations were of a standard pattern: traverses along most roads (almost exclusively earth at the time) in the moving Land Rover, recording altitude (by aneroid), landforms, vegetation and soil appearance every kilometre or wherever there appeared to be a change in soil type. When there was such a change, a stop was made to record a soil profile and site details; often this was from a road cutting, which were numerous and could be freshened up.

Sampling for analysis

On return to headquarters in Zomba, samples were sent for analysis at Bvumbwe. The site and traverse data were related to the photo-interpretation boundaries, which were simplified, removing any that could not be justified from field evidence, and transferring them to the base maps. The transfer was made not by photogrammetric methods, accurate but very slow, but by simply overlaying roads, rivers and other details visible on both maps and air photographs. The cartographer would then draw up the map, and the Soil Surveyor draw up the tabular legend, to be sent to the Directorate of Overseas Surveys for printing.

In the traverse notes, extensive use was made of the descriptors 'gently undulating plain' and 'very gently undulating plain', separated by a maximum valley-side slope of three degrees. Colleagues around the world subsequently adopted these descriptors, as GUPs and VGUPs.

Much of the above was standard procedure at the time, although with some differences from comparable reconnaissance surveys. The scale for map compilation, 1:250 000, was smaller than that found in many countries. Time did not permit the slow process of linear traverses on foot through the bush, but the relatively dense settlement of the country allowed reason- ably comprehensive coverage along earth roads. These methods allowed rapid progress, mapping an area about half the size of England in two and a half years. This said, it should be stated clearly that the maps of Malawi are less accurate than those of, say, Uganda or Ghana. There were places where air-photograph coverage was missing, and boundaries had to be filled in by interpolation. The Soil Series were defined more broadly than was the normal approved practice.

Another difference was the absence of a team effort. To echo the words of Jackson, one might have written: 'Soil Survey of Malawi – Survey- or-in-Charge: Anthony Young'. Coupled with this was the paucity of scientific contacts, which amounted to one visit by Herbert Greene from the Colonial Office, and one visit (on local leave) to the University of Salis- bury (Harare). There were no Departmentally funded journals. The survey undoubtedly suffered from this isolation.

Although the only officer at professional level, I take this opportunity to pay tribute to the unsung heroes of the golden age of surveys, the African staff at technical level. The team was small: Reginald Mwangala, Cartographer, later replaced by George Mandala; Elias Banda, Technical Assistant; Evans Gunda, a young man of very limited education who did a wonderful job in the field taking soil samples; and Khanyera, Driver, ex-King's African Rifles.

Mwangala was a fine example of staff development. Educated only to Standard 6 (two years short of British O-Level/GCSE), he was first sent on a six-months' course to Nairobi; shortly after I left, he was given a year at the Directorate of Overseas Surveys, UK, and returned as a professionally qualified cartographer, later to work for the Geological Survey.

In one important respect the methods used were an advance on many reconnaissance surveys. This was the linkage between soils and agricultural performance data, achieved by cooperation with an agronomist, Peter Brown, at the national agricultural research station, Chitedze (near Lilongwe). The soil surveyor had identified the Soil Series for all the sites for which there were experimental data, especially a national network of fertilizer and maize variety trials. The agronomist then assembled agronomic data, combined it where there were two or more experimental sites on similar soils, and produced crop, fertilizer and management recommendations for each Series. I believe the initiative for this came in the first instance from Brown; I took it up, and added other opportunities to link with agricultural data, for example Tobacco Marketing Board data.

Which soils grow the best crops? Agricultural show, Dedza District, Malawi;
Local farmers with Peter Goldsmith and the author's children

Quite why the official title of the work was 'Agro-Ecological Survey' is not clear. Trapnell's pre-war project in Zambia had been called Ecological

Survey, and most other countries, even though they almost invariably made comprehensive studies of land resources for agriculture, called them soil surveys. An *Agro-Ecological Survey of Southern Rhodesia* was produced in 1960, but I was unaware of this at the time. The Malawi study was effectively, like many so-called soil surveys at reconnaissance scales, a land systems study.

I resigned from Malawi in 1962, leaving the reconnaissance survey only two-thirds done. It was completed by Alan Stobbs, seconded from the Land Resources Division 1963-65. His survey of Southern Province is more detailed and probably more accurate, owing to the better maps and air photographs available by then, and his institutional backup. There were, however, regrettable delays in publication. The map for the Southern Region was produced only in 1971, six years after completion of the survey, and the memoir never appeared.

The results of the Agro-Ecological Survey were presented as a map in three sheets, for Northern, Central and Southern Malawi.[1] The mapping units were called Natural Regions, divided into Natural Areas. These were unorthodox titles, the choice of which I can only attribute to a sense of obligation to traditional geography of the time, but local users of the maps did not seem to find them strange. The Natural Regions should have been called Land Systems. The Natural Areas were certainly not detailed enough to be considered land facets, and should probably be considered Subsystems.

The detailed tabular keys were printed on the map sheets, and showed, for each Natural Region and Area, the standard material for surveys of natural resources: altitude, temperature, rainfall, soil parent materials (geology, etc.), landforms, vegetation, soils, and present land use. The soil classification employed is basically that of the *Soil Map of Africa* (1964), for example, the red soils are ferruginous, ferrallitic, etc., together with lists of Soil Series represented.

In the accompanying memoirs Part One, The Physical Environment, contains the standard material of soil surveys: geology and geomorphology, climate, vegetation, soils, agriculture, and the synthesis of these into natural regions. It is Part Two, Soil Series, which is distinctive. For each Series, the usual material is first given: classification, parent material, occurrence

---

[1]   The first is entitled, correctly at the time, Northern Nyasaland.

(Natural Regions and Areas), and a profile description with analytical data. This is followed by text of approximately equal length on Agronomic Data. An illustrative example is as follows.

SALIMA SERIES

Genetic group	Calcimorphic alluvial soil.
Vegetation	*Acacia-Adansonia-Hyphaene-Sterculia* cultivation savanna.
Occurrence	Salima Lake Shore Plain, mainly Natural Area 30c, also …
Morphology	A greyish brown alluvial soil with depositional bedding and imperfect site drainage; dark to very dark brown, mottled in depth; sandy loams and sandy clay loams …

AGRONOMIC DATA

Type sites	Agricultural Instructors' gardens at Pemba and Mtonga Markets, Ntakataka …
Potential	A soil of good potential for arable crops if well drained and weeded.
Nutrient status	Nitrogen is moderate to low, but phosphate and potash normally adequate
Agricultural characteristics	The profile is deep, but root room often limited to about 3 feet by a high wet-season water table. Easy to work when moist but gets hard when dry. Conservation measures should aim at getting excess water away quickly. Yields can be limited by weed competition, waterlogging, and nitrogen shortage. Unfertilized yields … maize 4-10 bags/acre … groundnuts … cotton (if sprayed) 800-1600 lb/acre. Prompt weeding is essential.
Response to fertilizers	(Data from Pemba 1959, Ntakataka 1958, Golomoti 1960.) Responses by maize to sulphate of ammonia 2-3 bags for the first and second increments of 100 lb/acre fertilizer applied … No recorded response to superphosphate.
Suitable crops	Maize (Mlonda or Mthenga cultivars); groundnuts (Early Runner), cotton, cassava, castor.

Information of this kind is given for the seventy-five Soil Series in Northern and Central Regions, and there are regional keys to their identification.

In a subsequent article, *Soil Survey Procedures in Land Development Planning* (1973) I expanded the view that surveyors should identify the soils at all sites where agronomic experimental work, or other agricultural performance data, were available. It is essentially the approach that Hardy followed in the West Indies.

How much use was made of this survey? This is always a difficult question to answer, the more so when the results became available just at the time of transition between Colonial rule and independence. The cynical will say that the maps made excellent wall decoration in government offices. On return visits over the ensuing forty years, government officers were very respectful when they learnt that I was part of 'Young and Brown', and commented on how useful the survey had been; but as has been said in a different context, 'They would, wouldn't they?'

Hence comments on how the survey was put to use must remain speculative. I do not think it likely that a planning committee ever waved a hand towards the map, commenting to the effect, 'Well, we know that flue-cured tobacco grows well in parts of the Kasungu Plain, so there must be potential to extend its cultivation over the wide area shown as Region 23.' It is more realistic to suppose that the Agro-Ecological Survey of 1959-65 subsequently served some or all of the following purposes:

- To provide guidance for more detailed surveys of proposed areas for development. From the 1980s it became common to development agencies to say, 'We don't need to do a soil survey, we'll make use of existing data.' This is a regrettable attitude, since reconnaissance surveys are a very inadequate basis of local development planning. Be that as it may, the Agro-Ecological Survey would have helped in first approximations of areas to be covered by development schemes. An example is found in project formulation missions by FAO in 1983 and 1985, of which a member of the team, Anthony Mitchell, writes, 'We made considerable use of the extensive soil and agro-ecological surveys (Young, Brown, Stobbs);

these were enormously valuable as they brought together so much information on soils ... and associated farming systems'.

- To form the basis of a national framework of soil types. The Soil Survey of Malawi continues to make local surveys for development purposes. The earlier reconnaissance study gave them a framework of soil series on which to base these.

- To link agronomic research with extension. The agronomic data, an example of which is given above, based on identified experimental sites and relevant to defined soil series, provides a means for extension staff, with advice from the research branch, to make recommendations for land management. The detailed, soil-specific, agronomic data assembled by collaboration between agronomist and soil surveyor is considerable. It makes a host of practical recommendations for land management, in terms understood by extension officers. It provides a framework for continuous updating of such recommendations, as new agronomic data become available. If the fullest possible use is not made of such a store of information, the fault does not lie with the surveyor!.

- Educational. Two very different kinds of development workers might benefit. The first are newly-qualified Malawian agricultural officers, extension and research, already familiar with their home region, to whom the survey would provide an introduction to the resources, potential and land management of the country as a whole. The second are expatriates on development projects, many of whom have no previous experience of the country and would seize upon maps and descriptions of their new assignments.

Land use

Fundamental to assessment of the agricultural development potential of any country is to compare the extent of cultivable land with the area already cultivated. In the 1950s most countries had land that was cultivable but not presently occupied, and hence this was the era of land settlement schemes. In later years, population increase led to spontaneous migration onto unused land. In Malawi this produced a situation in recent years, in which all sustainably cultivable land is occupied, together with substantial areas where there are serious hazards of erosion. In simple terms, the agricultural land of the country is 'full up'. If this were not the case, why would we find an average farm size of under 0.5 ha, or people feeding their families by cultivating slopes of 40%?

International statistics on land use are often inadequate and unreliable, in some cases almost ludicrously so. It is therefore unfortunate that systematic surveys of present land use in developing countries are rare. Malawi, however, has one. Alan Stobbs assembled air photographs for the period 1965-67, sampled each one randomly using pinpricks from templates, and recorded the land use at every sample point. The results are given as tables, showing the land use both for administrative districts and for the natural regions and areas of the agro-ecological survey. These could have been most valuable data, but for a number of limitations. First, they were never presented in the form of maps. Secondly, they were not published until twenty years later. Lastly, as the survey itself records, 'It had been the intention of the Malawi government to update these land use data every five years, by taking new air photography and repeating the sampling. Unfortunately this was not done.' This was indeed regrettable for repeated surveys, the later ones based on satellite imagery, could have led to a unique record.

Subsequent land resource survey

After independence in 1964, British expatriates continued to staff the Soil Survey for longer than in most countries. Anthony Mitchell, transferred from the post of Land Husbandry Officer, was in charge from 1968-75, and David Billing 1975-79. Early local surveyors were E.M. Ntokotha, Max Lowole and Nelson Nyandat, who returned from an MSc at Reading University to take charge of the Soil Survey Unit. They faced a difficult task. The Unit was under-funded, and sidelined by the expansion of the Land Husbandry Section. In 1965 Lowole produced a draft soil map of Malawi, based on compilation from existing sources (largely the Agro-Ecological Survey).

The staff continued to be active. A report, *Soil surveys 1971-72* assembles no fewer than eighteen of these: seven for irrigation projects, seven for general agricultural developments, and four classed as studies of specific problems. The last and longest of these, 'Ngabu cotton growing areas: reconnaissance soil survey,' attracted my attention as this was an area in which I worked on first arrival in the country. Reassuringly, the new survey recognizes there is an overlap with the area which I had mapped in 1960, and adopts several of its soil series names. The purpose of the new work was to investigate reasons for patchy growth of cotton, including the 'very serious increase in soil erosion, due to deterioration of soil structure under cultivation.'

In 1988-92 FAO conducted a Land Resources Evaluation Project of Malawi, with Robert Green as Chief Technical Officer. The maps are very much more detailed, and no doubt more accurate, that those of the Agro-Ecological Survey. When I remarked to a Malawian research officer that this must have rendered the earlier work obsolete, the reply was, 'Oh, those maps are too complex, we prefer to use yours.' It was, of course, highly desirable that after some thirty years, a new survey should have been conducted, and if there is a problem with complexity, then the fault lies with the presentation.

The head of the Soil Survey, Allan Chilimba, wrote in 2006 that he has recruited three new soil scientists from the University of Malawi, for whom they are seeking scholarships for postgraduate training, and that two detailed studies for irrigation projects had recently been completed.

Geology, forests, soil fertility, and population

Malawi has for long maintained a good record for research in all branches of natural resources. In 1922 Frank Dixey was appointed Government Geologist, and for the first five years had the whole country in his hands. The southern section of the African Rift Valley runs right through the country, dividing into two branches at the southern end of Lake Nyasa, then reuniting to become the Upper and Lower Shire valley. Dixey described the structure of this in a series of articles from 1926 onwards. At the same time he was carrying out mineral investigations, including coal in the north and bauxite on the summit plateau of Mlanje Mountain.

Unlike some geologists, Dixey did not shun geomorphology. *The Physiography of the Shire Valley, Nyasaland, and its Relation to Soils, Water Supply and Transport Routes* (1925) is a classic of its kind, and in a series of publications over the next thirty years he anticipated the work on erosion surfaces for which the South African Lester King was later to become well known.

In 1924 Dixey, J.B. Clements (Chief Forest Officer) and A.J.W. Hornby (Agricultural Chemist) travelled together, and wrote on the effects of vegetation clearance on climate, water supply and soil erosion. It would have been an inestimable privilege to have accompanied them.

In 1927 Dixey took the title of Director of the Geological Survey, justifying this by the appointment of three geologists by 1934. Mineral investigations took priority. During the Second World War there was again only one geologist, then in 1945 W.G.G. Cooper was appointed Director, and staffing and activities were built up. Systematic geological mapping began only at that time, and by 1955 one third of the country had been mapped. Cooper brought together achievements to date in *The Geology and Mineral Resources of Nyasaland* (1950, revised edition 1957).

Despite the variety of its geology and the presence of rift faulting, Malawi never found the 'golden egg' of mineral resources to support the economy. Coal was mined on a small scale, and limestone for cement manufacture on a larger scale. In 1960 I came across a one-man asbestos mine, with extraction in baskets from a hole. The country is unparalleled in the world for ring structures of carbonatites, but these have not proved a source of wealth.

There was an active Forest Research Institute, with a high quality of forest inventory in the Forest Reserves, and during the period under review, generally good protection of these against incursion. In 1970 the Commonwealth (now Oxford) Forestry Institute, Oxford, produced a comprehensive account of the evergreen forests.

Soil fertility decline has been taking place over many years, the primary cause being maize monoculture. In 1949 there was a drought and serious famine ensued. The Department of Agriculture Annual Report for that year contains the following:

> There are depressingly few signs that the Africans themselves appreciate the decline of fertility going on around them … Staff shortages prevent the problem being tackled on anything like an adequate scale.

The introduction of improved varieties, with cheap fertilizer, temporarily ameliorated the problem in the 1960s, but this was not to last. On a recent visit, my wife asked a petrol station attendant what was the main problem of agriculture in Malawi: 'It is the cost of fertilizer, Madam' he replied; and the miserable crops of maize to be seen all around showed that he was right.

On a return visit in 1999, forty years after I had first arrived in the country, I formed the view that population increase had checked or reversed all the advances which had been made over this period. The government now had no viable development options, other than the long-term measure of limiting population increase. As early as 1955 the Director of Agriculture, R.W. Kettlewell, had foreseen the problem; in a paper presented to the Legislative Council he wrote, 'Progress will be nullified unless Nyasaland's present rate of growth of population is substantially reduced.' At that time the population was three million, by 2017 it has passed seventeen million.[1] These conclusions may seem a long way away from soil survey, and yet land resource and population policy are integrally linked.

Land husbandry and conservation agriculture

The history of soil conservation, and its evolution into land husbandry and conservation agriculture, is a large and important subject which deserves a book to itself. It is set out briefly in Chapter 13. Malawi, however, played such an important role in the development of modern approaches that a summary is given here. The modern approach is now more usually referred to as conservation agriculture, but in view of the pioneering role played by staff based on Malawi I retain in this chapter the earlier term which was used there, land husbandry.

Soil conservation had for long been a concern of the Department of Agriculture. The country never adopted terracing, and the broad earth bunds found elsewhere were constructed only in a World Bank scheme, the Lilongwe Land Development Project (where they were not maintained). The main method advocated, and very widely adopted from pre-war days onwards, was contour ridging, simply hoeing the cultivation ridges required for maize along the contour, with uncultivated marker ridges at intervals. A Conservation Officer, Captain J.M. Howlett, was in post in 1942, and the

---

[1] Exceeding the population of its much larger neighbours, Zambia and Zimbabwe.

appointment of W.J. Badcock as Chief Conservation Officer in 1949 shows the existence of a team.

From the late 1950s, new approaches began to be developed. These are reflected in the changing titles of what was essentially the same post. In 1956, Garry Godden (South Africa) was called Soil Conservation Officer, and from the same year, Robert Green's post was Conservation and Extension Officer. Among their successors, Francis Shaxson held the title of Land Utilization Officer 1958-62, but in his second spell of service, 1968-76, he became Land Husbandry Officer. Malcolm Douglas held the latter title 1974-80. Shaxson and his colleagues produced *A Land Husbandry Manual* (1977), which includes an adaptation to local conditions of a system of land capability classification which originated in the US. This was widely used by the Conservation Service in the field-scale layout of local settlement schemes, although even in the locally-adapted form, the slope limits for arable use were at odds with the reality of the local situation.

The manner in which soil conservation came to be translated into land husbandry or conservation agriculture, integrating conservation with improved land management and production, is outlined in Chapter 13.

Recollections

When asking correspondents from around the globe to send details of their careers and achievements, I would add a request for unusual or bizarre experiences, so I must not personally shirk this obligation. Field travel on duty, familiar elsewhere as safari, in Malawi is invariably referred to by the Chi-Nyanja term *ulendo.*

My introduction to etiquette on *ulendo* came from George Jackson. When surveying near a village, you should set up camp in a compound in the centre of it, so that the children could come, stare, and report on the apparent innocence of your activities. If you camped in the bush, as reserved Englishmen might prefer, you could be engaging in witchcraft. The village headman would then send a present of a chicken, to which you responded by a present of money two or three times its value.

In connection with his work for the Boy Scouts, Jackson needed some camp-fire songs. His Chi-Nyanja version of 'The grand old Duke of York' is a model of translation, retaining the meaning and with perfect scansion. With a literal translation this went:

Kawinga wapita	Kawinga has gone
Wapita ku nkhondo	Has gone to war
Wapita kumwamba phiri	Has gone to the top of the hill
Wabwera munsi nso	Has returned to the bottom again
Pa nthawi pamwamba	At times on the top
Pa nthawi pamunsi	At times at the bottom
Pa nthawi ali pakati	At times at the bottom
Si mwamba, si munsi.	Not on the top, not at the bottom.

Lacking such proficiency in language skills, I composed an ecological verse. In the Malawian savanna there is an inconspicuous tree locally called *msolo* but having a very much longer botanical name. This inspired:

The Taxonomist's Reward

He went a-plant collecting, but came back on the dole-o,
All that he could find was the specimen *msolo*.
What was it eased his troubles, and left him feeling holier,
Pseudolachnostylis maprouneifolia.

The heat of the Lower Shire Valley is formidable. If one sympathised with local expatriates, they would remark that in July and August (midwinter) it was really quite pleasant. On descending the Rift Valley scarp, one arrived at the River Shire. This is now crossed by a bridge (inevitably Kamuzu Bridge), but in earlier days it was crossed by a small ferry with an energy-efficient method of propulsion. A metal cable connected the two banks. The ferry was attached to this by two ropes, front and back, the lengths of which could be altered by winches. The ferry was set at an angle of 45 degrees, when the current would carry it effortlessly across. Adjusting the ropes to the opposite angle powered the return journey. The local District Commissioner would bring a deck chair and sit on the ferry in the evening, smoking his pipe as he travelled to and fro.

Internal self-government was granted in 1962. Following elections, Dr Banda became Prime Minister, later to be Life President. In the first instance he also took the Natural Resources portfolio, and as such visited the Department of Agriculture. This included the office of the Soil

Surveyor, in which were laid out air photographs of the Scottish Mission at Livingstonia, in the Northern Province, where he had received his early education. Thus in the same tour of service, I had indirectly assisted in putting Dr Banda in jail during the State of Emergency, and subsequently shaken his hand as my Minister of Natural Resources.

Distinguished visitor: Dr H. Kamuzu Banda, shortly to become
Life President of Malawi, being shown air photographs of Livingstonia
Mission, where he received his early education; Anthony Young centre

Forestry Departments are mostly allotted land unsuited to agriculture, and in Malawi this meant the high-altitude plateaux: the Nyika and Mlanje (which appear in Laurens van der Post's *Venture to the Interior*), the Vipya, and Dedza and Zomba Mountains. When in 1973 I returned to Malawi with a research associate, Peter Goldsmith, to carry out a pilot study of land evaluation, we were kindly allowed to live in a Forestry Department house on the slopes of Dedza Mountain. Naturally, everything was constructed from wood products. One day returning from fieldwork a smell of burning directed us to the bathroom. Looking up, a large circular hole had developed in the chipboard ceiling and was smouldering outwards, about to reach the wooden slats. I must have been shaving before first light using an 'Aladdin' lamp with its paraffin wick and tall glass cylinder, and placed it on a shelf whence the heat rose upwards. Dowsing this ring of fire in the nick of time saved us from, at the very least, the embarrassing situation of having burnt down our host's house, and at worst, from starting a forest fire.

Self-restraint must limit such recollections. On a more serious note I would add that my experience was common to all Colonial surveyors, namely an invariable welcome and kindness from local people throughout the country. Even during the Emergency, it was rare to encounter hostility.

During the initial tour of service, the normal passage for officers was changed from sea to air (with the consequence that the length of leave, after three years, was reduced from eight to six months). I did not avail myself of this new means of transport but instead, on 24 February 1962, set off in a Volkswagen Beetle that had been shipped out at the start of the tour, driven on earth roads, and already had 110 000 kilometres on the clock. My wife and I drove home through Malawi, Tanzania, Uganda, Sudan, Egypt, Libya, Tunisia, and thence to Sicily and on to Britain, the journey taking two months. The three and a half years of largely self-guided fieldwork in Malawi was to act as a springboard for a career in land resource science, an experience held in common with many surveyors in other countries.

9

THE WEST INDIES AND CENTRAL AMERICA

The West Indies

ENVIRONMENTAL AWARENESS AROSE early in the West Indies. All forms of economy required clearance of the forests which clothed the islands. It became apparent to many observers that this was followed by reduction in the flow of streams, and to some that it led to erosion or a rapid loss of the original soil fertility. They also thought, incorrectly, that forest clearance would reduce the rainfall.

In Barbados, the Governor of the colony was made aware of the effects of plantation development as early as the 1660s. After thirty years of cultivation, the land 'renders not by two-thirds its former production per acre.' During heavy rain the soil would 'run away', and landslides took place on deforested hillsides. The St Vincent Assembly in January 1790 discussed soil erosion and the causes of gullying, and gazetted a forest reserve. Conservation and diffusion of biological resources received attention through the establishment of a network of botanic gardens in the late eighteenth century, including on St Vincent and St Helena, a movement spearheaded by Sir Joseph Banks.

By the mid-twentieth century large parts of most West Indian colonies had been cleared for agriculture, particularly perennial plantation crops such as bananas and cocoa. Small farmers displaced by plantations had to retreat to the hills to find land for food crops, inevitably leading to erosion. Some well-meaning efforts to combat this were made by the World Bank, unfortunately led by a conservationist of Taiwanese origin who mistakenly supposed that terraces, successful in parts of the Far East, could be introduced to Jamaica. Through lack of maintenance almost all of these were abandoned. The approach of conservation through biological means, maintaining a complete ground cover under plantation crops, was more effective. Erosion and other problems, notably those of land tenure, and the dependence of the economies on agricultural exports called for a rationalisation of land use and management, for which a knowledge of land resources was an essential basis.

Institutions

The greatest contribution to the success of soil science in the West Indies came from the continuity of its institutions. The Imperial College of Tropical Agriculture (ICTA) was established at St Augustine, Trinidad, in 1921 with soil science, under the leadership of Fred Hardy, as one of its founding disciplines. Its purpose was to train agriculturalists for Colonial Service throughout the British Empire. In research, it gave particular attention to problems of the Caribbean region.

The University of the West Indies (UWI) was founded in 1948, initially as a college of the University of London. At the political level there was an abortive Federation of the West Indies 1958-62, after which most of the territories became independent, but fortunately this did not lead to the fragmentation of research and teaching. In 1962, UWI became an independent university with three campuses: Mona, the original site in Jamaica, housed medical sciences, Cavehill, in Barbados, law, whilst the former ICTA became the St Augustine campus, taking on engineering as well as agriculture. Today the University is an autonomous institution serving fifteen countries, all but one, Belize, island territories.

In 1954 teaching and research in agriculture at St Augustine were separated, with a Regional Research Centre (RRC) taking over research whilst the University of the West Indies was responsible for teaching. When ICTA became the University's Faculty of Agriculture in 1963, the RRC was established as a research centre within it.

To this exceptional degree of institutional continuity was added a pooling of personal experience. Fred Hardy had masterminded soil science from 1921 to his 'retirement' in 1954. Meanwhile, Alun Jones had become Reader in Soil Science, and spent his first year attending Hardy's lectures. He was later given the option of heading the teaching or research side, and chose to become Head of Soils Research at the RRC, whilst Harry Vine was appointed as Hardy's successor in the University. Jones remained for six years, to be followed by John Coulter as Head of Agronomic Research at the RRC 1959-63. In 1969, Nazeer Ahmad became the first West Indian Professor of Soil Science.

The Cocoa Research Scheme

Hardy had always insisted on close integration between soil survey and agriculture; some of his early 'Grey Book' surveys had titles like *The cocoa Soils of Tobago* (1931). In the late 1920s the Lovat Committee of the Colonial Office envisaged the formation of a chain of agricultural research centres to serve different regions of the tropics. These would undertake long-term research of a kind not normally done by national Departments of Agriculture (a precursor of the chain of international agricultural research centres established in the 1950s). The ICTA Professor of Botany drew up a memorandum on research needs for cocoa, Trinidad's largest export crop, and in 1928 the Principal, Sir Geoffrey Evans, called a meeting of government officers, planters and merchants who enthusiastically supported the proposal.

The Colonial Office gave their approval and in 1930 the Cocoa Research Scheme was launched. Five governments with cocoa exports contributed to its financial support: Trinidad, Sri Lanka, Grenada, Ghana and Nigeria. Sri Lanka, for example, promised £100 capital and £50 per year for five years. To this was added commercial sponsorship from, not surprisingly, the companies of Cadbury, Fry and Rowntree. The Colonial Office topped up these contributions to bring the recurrent budget to £4655 per annum (about £150 000 today).

Surveys of cocoa-growing districts were conducted, based on soil profiles at 'good' and 'bad' sites selected by estate managers in Trinidad, Tobago and Grenada, and on a central research station, good and bad plots were monitored for soil ecological factors. This scheme was successful, and by 1970 over ninety papers had been published. It later became the Cocoa Research Unit within the University.

Soil surveyors often do not pay enough attention to linking soil factors with crop yields. Anyone who complains of this can go to a paper, 'The effects of soil type and age [of tree] on yield'. Published in 1932, it gives immensely detailed data on yields for estates on three soil types, listed tree by tree, annually for 1910-36. The results were obtained before statistics were extensively applied to agriculture, but these data would repay analysis.

The 'Green Book' surveys

The start of soil survey by Fred Hardy has been described in Chapter 2. Thirteen reports, the 'Grey Books', were published in 1922-35, plus three in 1947. These were focused on crop relations of soils, supported only by sketch maps. Funding support for the more substantial post-war survey effort came from the Colonial Development and Welfare Fund, with early staffing from the Colonial Pool of Soil Surveyors, later by secondment from the Land Resources Division. Staff from the Colonial Pool became available from 1951, and fieldwork began in Jamaica and St Vincent in 1952. Altogether twenty-four surveyors were involved in the second round of surveys. A number of these, including David Lang, Douglas Carroll and Ian Hill, continued work in other countries with the Land Resources Division. William Panton moved to Malaya where he pioneered soil survey and remained for the rest of his career.

Unlike the national reconnaissance studies of nearly all other countries, the small areas involved allowed the West Indian surveys to be undertaken at detailed scales, mainly 1:25 000. Work proceeded concurrently in Jamaica and the smaller islands, with the results published in reports known, again from their covers, as the 'Green Books'. The results for the island territories were published 1958-1967, and for Jamaica 1961-1971. The administrative divisions of Jamaica are based on the English countryside with three counties, Cornwall, Middlesex and Surrey. These are divided into thirteen parishes, mostly with names such as St Catherine and St Mary but also Westmoreland, Manchester and Hanover. The parishes were of convenient size to provide the basis for soil surveys on the same scale as those of the islands. In Trinidad some even more detailed surveys were made, with a training element, at 1:5000, a scale not often found outside irrigation schemes.

The twenty-six 'Green Books' are made up of thirteen for the Jamaican parishes, ten for the island territories, and three for sample areas of Guyana. Under the guiding influence of Alun Jones, all follow a similar format. The

1958 survey of St Vincent may be taken for illustration. The surveyor and lead author is Philip Watson, later to become Professor of Soil Science at the University of Zimbabwe, Harare,[1] and subsequently at Venda University in South Africa. There are three parts to the report: factors affecting land use; soils; and land capability and recommendations. The most substantial element in Part I is 'The people and their agriculture', an historical account of agriculture and land tenure from the earliest settlement, followed by current land use and trends, soil conservation and soil management. The farmers do not like bench terracing, but the alternative of grass strips was not always effective. Features characteristic of the West Indies are found: cultivation of steep slopes, a heavily skewed distribution of farm sizes, and dependence on a small number of export crops. In the case of St Vincent there are 7000 farms of less than 4 hectares, 240 of intermediate size and sixty-seven of 40 to over 400 hectares. The island is the world's principal supplier of arrowroot, which is its main export.

Part II on soils supports the map, in the case of St Vincent at 1:20 000 scale. Mapping is at the level of soil series (e.g. Bellevue Sandy Loam), grouped into descriptive classes originating from Hardy's pre-war survey, such as yellow earths, or recent volcanic ash soils. There is a section on trace element deficiencies, including results from field trials. Part III begins by assigning the soil series to the US land capability classes, I-VII, which were in common use at the time. This is followed by a short discussion of problems of land tenure and fragmentation, and suggested lines of research. Appendices give the essential building blocks of survey reports, profile descriptions with analytical data and, for each soil type, present land use and recommendations.

The nature of the 'Green Book' maps arose from a chance encounter. In 1955 Alun Jones travelled home on leave with E.M. Chenery, who had completed the initial survey of Trinidad and was carrying the highly colourful map. It had, he said, taken eight years to complete the printing, largely because this had been handled from London. As a result of this encounter, Jones explored the possibility of getting the printing done locally. The *Trinidad Guardian* agreed to undertake this task at very modest cost, but their equipment at the time was capable of handling only four colours. He decided that the maps were needed sooner by the farming

---

[1]   At that time, the University of Southern Rhodesia, Salisbury.

community, rather than later as cartographic showpieces. Drafts for the cartography were tailored to this requirement, hence the maps are rather modest affairs.

Subsequent to work by the RRC, the Bahamas were covered in two surveys by the Land Resources Division, an inventory of the pine forests (1974) and an overall appraisal of land resources (1977). A well-printed and presented *Land Capability Survey of Trinidad and Tobago* (1965-1966) was produced with the support of four commercial sponsors, three in oil and one in sugar.

The 'Green Book' surveys are now some fifty years old, but retain much of their value. All in all, they are model examples of good soil survey and the contribution that it can make, given support from research and extension services, to agricultural development. The West Indies share with Uganda the sequence of having had a land resource survey task conceived, executed, and successfully completed.

Belize (British Honduras)

The colonies on the American mainland, Belize and Guyana, had different survey histories. Belize is only twice the size of Jamaica and might have been included in the 'Green Book' surveys but did not fall within the mandate of the Regional Research Centre. In 1940-41 there had been two reports by C.F. Charter, one on sugar cane soils, the other a descriptive account of soil types. The main reconnaissance survey was carried out as a Colonial Development and Welfare Scheme in 1952-54. Subsequently, the Land Resources Division carried out a forest inventory in 1974 and a systematic land resource assessment, led by Bruce King, in 1992.

The survey of 1952-54, *Land Use in British Honduras*, is an astonishing piece of work. Over 200 000 words long, it was not published until 1959, half the copies were lost in hurricane flooding, and it is now hard to obtain. It was led by Charles Wright, D.H. Romney and Ron Arbuckle (New Zealand). The conventional parts of a soil survey report, environment and soil descriptions, occupy only 10% and 20% respectively of the report. After the usual accounts of climate, geology, landforms and vegetation, the greater part of the environment section is given over to an account of the people — Maya, Carib, Creole, and 'East Indians' (immigrants from India) — and their farming practices, particularly *milpa* (shifting cultivation).

There are maps of soils, vegetation, and potential land use at 1:250 000, based on field survey at 1:40 000; the latter, 'a fairly detailed soil survey of the whole country ... was left in Belize where it should be accessible to all who wish to consult it.' The soil descriptions are immensely detailed: the extended legend to the map, together with 'Supplementary Notes' on the soils, occupies sixty pages, even though, 'we have pruned it as much as we dare.'

An old-style, conventional, survey would have stopped here. In contrast, this report now moves into Part III, 'Developing the soil resources of British Honduras', which makes up two thirds of the text. Region by region, 'Suggestions for improved land use' are made, divided into immediate projects and long-term projects. These are by no means simply physical land evaluation, but take into account the needs and circumstances of the farmers:

> We have also included our comments upon the people, their way of life and their reactions to the new ideas we laid before them as we sat and talked in the cool of many an evening. People are the real stuff of development.

Also discussed are problems of land tenure, the lack of legislation to safe-guard wise land use [this would now be called sustainability], and the need for better marketing facilities.

There had been nothing of this kind before, and it was not until the integ-rated development feasibility surveys of the Land Resources Division in the mid-1970s that the like was to be found again. Old-style, conventional, soil surveys end with a somewhat perfunctory chapter on potential land use, whilst development studies centred upon economics often overlook the problems of land variability which are faced by farmers. In this study, trans-lation of the results into practical development potential was achieved without sacrificing the details of soil and environmental variability obtained through the field survey.

This difference in focus and emphasis was not present in the terms of refer-ence, which specified a stocktaking of the soil resources of the country with a view to determining the 'correct' land use. It appears to have arisen

spontaneously within the field team, without calling in economists or sociologists:

> In the course of eighteen months we were mainly gathering inform-
> ation about the soils and their potentialities, but at the finish we
> found we had accumulated a great deal of miscellaneous informa-
> tion ... We have included much of this material in the report ...
> The objective, after all, is to create new conditions ... in which a
> segment of the human race can enjoy the maximum prosperity in
> any manner it may desire.

Charles Wright became so dedicated to the country that in retirement he
chose to live there, in the forest and among its people. *Land Use in British
Honduras* has not been given the attention it deserves.

Guyana (British Guiana)

Guyana occupies 215 000 square kilometres, ten times the size of Jamaica,
so the scale used for West Indian surveys was out of the question. As head
of the RRC, Alun Jones received a request from the British government that
he should team up with Charles Wright to do a reconnaissance survey of
the whole country. 'This was to be done on a budget, if I remember
correctly, of $1500, and there was a last-minute telegram warning us to have
due regard to costs.' Jones and Wright were among the first to use aircraft
for transport and observation during soil survey:

> We started off by hiring a DC3 and flew it for ten days, keeping it
> in the bush for three nights. Then we hired a Grumman Goose to
> fly the northern part of the country where river landings were
> demanded because of the more or less continuous forest cover.
> Even then there was still some cash left.

The next stage involved cooperation between the RRC and the US Foreign
Aid Office, USAID. Alun Jones' counterpart was Gerry Bourbeau from
Cornell University, assisted by the son of Roy Simonson, a famous name in
US soil science. The decision was taken that the US team would tackle the
coastal swamps (to the great relief of Jones and Wright) and RRC would
start on the interior.

Subsequently, reconnaissance and semi-detailed surveys were completed by FAO in 1961-66, with a multi-national staffing including Roy Suggett from England, Clyde Applewhite from Trinidad, and Robert Brinkman from the Netherlands, who was later to become head of FAO's Soils Service.

Recollections

Gerry Bourbeau and Alun Jones sat down and conferred at to what equipment would be needed. Alun recalls:

> I vividly remember him putting down on paper the first item, a refrigerator! I said you cannot possibly transport a frig around the swamp lands. His response was that without a frig he would be unable to recruit a single American. Realising the difficulty of movement, it had been decided to use a helicopter to drop survey teams at designated spots. I believe it was the first time that use of a helicopter for soil surveys had been attempted. Anyway, the whole thing had to be dropped because the 'copter got stuck in the mud and couldn't take off. I do not know to this day whether it was ever recovered.

A large part of the interior of Guyana consists of the Rupununi savannas, a very ancient landscape with old, highly-weathered, soils. The same plateau runs across into Venezuela and Brazil. The country is so nearly level that the watershed which is supposed to mark the boundary is barely discernible. A number of intrepid ranchers struggle to run a cattle industry in this remote and hostile environment, the first of them a Scot who became too ill to travel, was left to die by the Essequibo River, and nursed back to health by local Indians.

One of these ranchers was Tiny McTurk, who acted as host to Alun Jones and his colleagues:

> Tiny was visited a year later by David Attenborough, sponsored by the Royal Zoological Society and the BBC. Out of this trip he published *Zoo Quest to Guiana*. When my wife read it she wanted to know why I hadn't written it because all the anecdotal detail in it she had heard previously from me!

I gladly acknowledge that much of the anecdotal material in this chapter originates in a copious set of letters from Jones.

10

SOUTH ASIA

India

India before independence

FOR INDIA, THE STORY is very different. In the first place, the sequence of dates and types of survey contrast with the sequence of events found in the rest of the Commonwealth. Secondly, most of the early work was carried out by staff of local origin.

'Agriculturalists in ancient India (3250 BC-AD 1200) were quite conscious of the nature of the soil and its relation to the production of specific crops.' Thus begins an account, *Soil Science in India*, written to celebrate the Golden Jubilee of the Indian Society of Soil Science in 1984. It goes on to give some names for soil types originating in those times: *jangala*, barren land, with scanty bushes; *anupa*, moist land, marshy with forest cover; and *sadharana*, ordinary land, the cultivated areas. Further distinctions were made on the basis of soil colour — it was known that the best soil was dark in colour, being rich in organic matter — and whether it was sweet, sour, bitter or pungent in taste (i.e. salinity and alkalinity). 'Elaborate injunctions are found in the *Atharva-Veda* (321-186 BC) ... regarding the use of fertilizers, animal excreta ... and various kinds of mixture decoctions.' Other injunctions predating AD 1000 make good sense in terms of modern science. For improved tree growth, dig a hole around it, burn the soil, and add bones and manure [potash aids tree growth]. A dung heap should be left undisturbed up to the month of *Magha*, i.e. for ten months [to lower the carbon-nitrogen ratio]. The use of leaves as compost, and growing crops beneath the canopy of certain trees, was extensive [agroforestry, a word not coined until 1978].

During the period of the Mughal Empire (established 1526), the Emperor Akbar classified lands according to the relative number of years cultivated and under fallow, from continuous cultivation to land fallowed for five years or more, fixing a different revenue to be paid by each. The *zamindar* or collector, as representative of the Emperor, was enjoined to be a friend of the agriculturalist, to encourage him to bring wasteland into cultivation [thereby, of course, increasing revenue] and, 'to assist the needy husbandman with advances of money'. This period also saw the construction of irrigation schemes, systems of canals fed either by offtake from rivers or by construction of 'tanks', small reservoirs. A good deal of ad hoc local survey would have been necessary to ensure that water reached potentially fertile land.

The period of British influence began with the trading posts of East India Company (from 1746) and was widely established by the middle of the nineteenth century. Land classification for purposes of the collection of revenue continued. The classes notably included *banjar* or 'culturable wasteland', the equivalent of today's assessment of land as cultivable but not presently cultivated.

Modern soil science in India originated from a Famine Commission of 1880, which called on the services of an adviser, Dr J.A. Voelcker from England, who wrote a *Report on the Improvement of Indian Agriculture*. The recommendations of this took seven years to materialise, at the end of which the adviser left the country — a sequence of events familiar today. Research in agriculture received a boost on the appointment of Lord Curzon as Viceroy. On his advice, the British Cabinet approved the establishment of the Imperial Institute of Agricultural Research, initially at Dehra Dun, then Pusa. Laying the foundation stone, Curzon observed that, 'Pusa had its natural beauties with a soil suitable for growing wide varieties of crops and ... suited to the combined objects we had in view.'

Among the founding staff appointed to the new institute in 1892 was a British agricultural chemist, Dr J.W. Leather. His contribution is described in a review of 1984 written by a group of Indian scientists headed by A.N. Ghosh:

> Dr Leather can rightly be regarded as the Father of Soil Science and Agricultural Chemistry in India ... Keeping in view the fact that

only a handful of workers were competent and available for making an inventory of the vast soil resources in the country ... he devised an ingenious mode of characterizing the soils into four major groups or associations viz. alluvial, black, red and laterite.

This classification, valid as far as it goes but highly simplified, was not to be replaced until the 1960s. 'An institution by himself,' the review continues, 'it is difficult to comprehend all the monumental contributions of Dr Leather.' The most important of these was to train staff who were to become the leading soil scientists in the country over succeeding decades. Leather left India in 1916, as a member of the British Volunteer Force in the First World War.

A few local resource surveys were carried out from the 1920s onward, in particular for irrigation projects. However, systematic national soil survey received a setback. The Royal Commission on Agriculture, in its report of 1928, observed that since the main classes of soils with their approximate distribution had been established, little further would be gained by soil survey, except local surveys for specific purposes. The first national soil map was prepared in 1932 by a Russian, Mrs Z.Y. Shokal'skaya, unique in that it is believed she never set foot in the country.

India since independence

Following the separation and independence of India and Pakistan in 1947, activity at national level resumed with the formation of the [first] All India Soil Survey Scheme, headed by S.P. Raychaudhuri. This received a stimulus when the Indian government invited A.V. Stewart of the Macaulay Institute, Aberdeen, to advise on reorienting studies on the fertility of soils. After visiting different States and meeting their agricultural chemists, he recommended that along with fertilizer experiments in farmers' fields, in a country so large and varied a survey of soil-climatic zones should be conducted so that a correlation could be made between soil types and crop yields. A further advisory visit was made by a member of the US Soil Survey, F.F. Raickens.

Much of the work of the first national scheme consisted in assembly and collation of information from the States of India. The results were presented in the *Final Report of the All India Soil Survey Scheme* (1953), later

expanded into book form as *Soils of India* (1963). The early four-fold classification by colour was still being used, extended by the addition of hill soils,
desert, saline, alkaline and marshy soils. The maps of each State are highly
generalized, most being based on reported information rather than survey.
About four soil types are typically shown for each State, although for
Manipur, near the border with Burma and for which there was very little
information, the map shows exclusively 'red soil'. The information from
States was combined into a very small scale national soil map.

As was commonly the case with early national soil maps, the most
important function of this was to show what was not known, as a stimulus
for further research. The All India Soil and Land Use Survey Scheme
(AISLUS) was established in 1956, again under the direction of S.P.
Raychaudhuri, with headquarters in New Delhi and four regional centres. In
addition, the Central Water and Power Commission carried out soil surveys
for irrigation schemes. In 1969 the AISLUS bifurcated. Part retained its
original title and remained with the Department of Agriculture, with
responsibilities for conservation and development surveys. The research
aspect, including soil correlation and the national reconnaissance survey at
1:1 000 000 scale, was taken over by the National Bureau of Soil Survey and
Land Use Planning at Nagpur, Maharashtra.

It was at this time that survey in India, which had hitherto trailed behind
international advances in soil classification and survey methods, caught up
with modern developments. A soil survey manual, based on the USA-FAO
standard methods, was published in 1970, and the FAO system of agro-ecological zones was applied to the country. India never lost the close links
between survey and field experimentation which had all along characterized
its activities.

By 1967, ten years after establishment of the new organization, some 42
million hectares, 13% of this vast country, had been covered by reconnaissance survey; by 1983 these figures had reached 115 million hectares or over
one third of the country. In 1965, a draft national soil map at 1:3 000 000
scale was presented to a soil correlation meeting in New Delhi. Three years
later this had been revised and converted into FAO soil units, to form the
country's contribution to the FAO-Unesco *Soil Map of the World*. Seven
contributors to this map are listed, all of local origin, among them the

senior figure of S.P. Raychaudhuri. Other surveyors of international repute by this time included J.S. Kanwar, J.L. Sehgal, and S.V. Govinda Rajan.

The sheer magnitude of agricultural research in India is striking. The Indian Council of Agricultural Research lists forty-three institutes, seventeen research centres and five bureaux, and this is for national-level centres only. Among these, the Central Soil Salinity Research Institute and the Central Arid Zone Research Institute are major world centres. In 1998 there were eighty-five national field soil survey units and a further 300 at State level. In one respect, India leads the world. The Indian Institute of Soil Science, Bhopal, Madhya Pradesh, was established in 1988 to conduct basic and strategic (basic) research in soils. This recognition of the importance of strategic research is in contrast with the demise of the corresponding body at international level, the International Board for Soil Research and Management (IBSRAM), which ended its sixteen years of operation in 2001.

This organizational structure may appear complex, but it is justifiable by the sheer size of the country. With a population well past the 1.3 billion mark,[1] about three-quarters of the people in the Commonwealth, and an average per capita land holding of 0.15 hectares, it is well that India has a large and established soil and land resource survey organization. In 1982 the International Soil Science Congress was held in New Delhi, only the second time it had been hosted by a developing country.

Satya Prasad Raychaudhuri (1904-1986)

For nearly twenty years, 1944-61, soil survey in India was led by S.P. Raychaudhuri. To the outside world, Indian soil science and Raychaudhuri were practically synonymous. Born in 1904 to a *zamindar* (landlord) family in West Bengal, he was an outstanding student, first at St Xavier's College, Calcutta and subsequently at Calcutta University. In the matriculation examination of the latter he secured the highest marks of all in his year in mathematics, and was awarded a scholarship of 10 rupees a month. Inspired by J.N. Mukherjee, he chose chemistry as his main subject, continuing to obtain academic honours, including a DSc at Calcutta for work on the electrochemistry of soil colloids. This was not to be his only academic distinction, subsequently being awarded PhD and DSc degrees from the University of London.

---

[1]   2016, post-independence India.

The turning point in Raychaudhuri's life was a scholarship to Rothamsted Experimental Station in the UK, where he worked on red soils of India under the guidance of two notable scientists, B.A. Keen and E.M. Crowther. His research into these soils continued during his first post, 1937-44, at Dacca University (now Dhakar, in Bangladesh). Although his work was laboratory-based, he became much concerned with soil fertility as it affected farmers. Like so many soil scientists who began as chemists, he must also have acquired during this period an appreciation of soils in the field.

This early career had prepared him to take the leadership, at the age of forty, of soil survey in India. On the foundation of the All India Soil Survey Scheme he was appointed its Director. This put him in a position to advocate the value of soil survey with the Indian Council of Agricultural Research (ICAR), a task in which he was notably successful. From 1949-58 he was Head of the Division of Agricultural Chemistry at ICAR, and when in 1958 the survey became the All India Soil and Land Use Survey Organiz-ation he again became its founding Chief, until his retirement in 1961.

Within India, he was the guiding spirit of the Indian Society of Soil Science, as Foundation Member (1934), Secretary (1952-59), and President (1960-61). Among international development institutions, as soon as India was mentioned Raychaudhuri would be the first call. When he retired from offi-cial duties he became still more available to the world at large, remaining active into his eighties.

Observers in the West admire how in India, men with outwardly quiet and unassuming personalities can attain the highest office. To convey the flavour of this, and give some idea of the respect in which Raychaudhuri was regarded by his peers, I cannot do better than quote from his obituary:

> Dr Raychaudhuri led a very peaceful life, in the family and in the office. His amiable disposition, unassuming nature and simple life style commanded admiration and respect. We never found him to lose his temper under adverse circumstances. His helpful attitude was remarkable, never saying 'no' to anybody. One must also mention his unflinching faith in science, and his sincerity and conscientiousness. He was so youthful that at scientific gatherings he would be addressed by very senior colleagues as 'young Raychaudhuri'. None of us could think he would leave us, until in

1986 his mortal frame was consigned to flames and this doyen of soil science left for his heavenly abode.

Pakistan (West Pakistan)

Survey with armed guard, near the northern frontier of Pakistan

For the part of former India which became West Pakistan the earliest systematic account dates from pre-independence, *Soils of the Punjab* (1929). It consists of a collection of soil analyses, noting that, 'no comprehensive survey of these soils has been undertaken'. A report on erosion in the Punjab was made by Sir Harold Glover in 1946, including many photographs of catastrophic erosion in the hill areas. His words on the duties of collectors, or district officers, deserve quotation:

> The officer entrusted with the duty of realizing land revenue is not a mere rent collector ... His position is rather that of the steward of a great landowner ... In particular, the Collector must do everything in his power to conserve the soils in his District and to maintain its fertility ... The aim of land policy is the true symbiosis, or permanent association, of man, his animals and the land.

What better expression could one find of the concept now called sustainability?

More than half Pakistan is hilly to mountainous, part of the Himalayan chain. Most of its people live on the alluvial plains of the Indus and its tributaries, the Punjab ('five rivers') or Upper Indus, and the Sind or Lower Indus. These plains are arid to semi-arid, with agriculture largely dependent on irrigation, and this was the reason for most survey activity.

Large-scale irrigation: constructing concrete liner tubes
for the Tarbela Dam on the Indus, the largest earth-filled
dam on the world, completed in 1976

Traditional irrigation methods were many, varied and ingenious, but with limitations imposed by the small height of lift and distance from the source of water. From the early twentieth century these were extended through construction of canal-based schemes. A map of 1931 shows that the major canals for the Punjab, each running along the *doabs* (interfluves) between tributaries, were already in existence. The Sukkur Barrage in Sind, for stabilizing the water level at the offtake, was completed in 1932. There are now three major reservoirs, the largest formed behind the Tarbela Dam on the main Indus. There are numerous barrages and a canal system of total length 50 000 km, irrigating 16 million hectares. Some idea of the scale of this system is conveyed by the names of successive stages of canal bifurcation: main canal, branch canal, major distributary, minor distributary, and water-course; from these last the water passes into farmers' land through a *nakkar* or flume, below which a further system of channels distributes water through the fields. Flying over such areas, the rectangular layout of fields gives an appearance like graph paper.

Canal headworks: from the River Indus to the Major Canal

Minor Canal

From Minor Canal to farmer's land:
measuring water flow through a nakkar

The end point: water reaches a cotton crop via a field channel

To construct such schemes requires engineering survey of a high degree of accuracy, in order to achieve a gentle but regular channel slope. There must also have been surveys of the land under 'command', that which lies below the level of a source of water. Some early surveys may survive in the files of government departments and engineering companies.

Post-war activity, however, was dominated by the problem of rising water tables and salinization. Saline patches had begun to appear during the war years, and by the 1950s had spread to render large areas of land sterile. Reclamation of these, through Special Conservation and Reclamation Projects (SCARP), was highly expensive.

Following independence and separation from India in 1947, the Soil Survey of Pakistan had to be newly constituted. Much of survey work was done by consultant companies funded by aid projects. The Photographic Survey Corporation of Toronto, working under a Canada-Pakistan Colombo Plan project, completed a reconnaissance survey of the Indus Plains in 1958. As consultant surveys go the report from this is short, one volume of 400 pages. There are two maps at 1:253 440 scale, one of landforms and soils, the second of land use, each of twenty-five sheets. These must have been produced largely from air photograph interpretation, with only limited ground survey.

Three years later in 1961, Huntings completed a Mangla Watershed Management Study. This was the forerunner of field surveys on an immense scale, the Lower Indus and Upper Indus projects, conducted jointly by Huntings and the engineering consultancy company, Sir Murdoch Macdonald. The multi-volume reports of these studies cover hydrology, soils, population and agriculture, including farm system studies. The atlas to the Lower Indus Report has a wide range of maps at 1:750 000 scale. Other consultancy surveys followed under the aegis of the World Bank.

An FAO soil survey of Pakistan was begun in 1959, initially covering both West and East Pakistan as they then were. The report for West Pakistan appeared in 1971. In the same year, East Pakistan broke away as Bangladesh, where survey work continued. As in most projects by this time, international staff from FAO worked alongside Pakistani soil scientists.

Recollections

I briefly worked for Huntings on the Upper Indus project in 1964, and from this experience witnessed a notable example of staff development. Part of our task was to recruit graduates from Lyallpur Agricultural University, who were to live in villages making farm system studies. Not all were of high quality, but one stood out from the rest. Many years later, at the welcoming cocktail party of some international conference, I was greeted with, 'You will not remember me, Professor Young, but you gave me my first job.' On being told which this was I asked what he was doing now, to receive the reply, 'Director of Lyallpur Agricultural Research Station', the leading station in the country.

Punjab village surveys

An extraordinary series of early village surveys were carried out in the Punjab[1], conducted by the Board of Economic Enquiry, Punjab, and published between 1931 and 1938. It was intended to survey one village in each of the twenty-nine districts, although only ten were published. The exercise was carried out as background to a land consolidation programme. 'It may be held that the information printed in the following pages is too detailed' states the Preface. 'To this the reply is that [it] is to answer the question: How do people live on the small holdings in a congested district like Jullundur?' The village maps indicate boundaries of every land holding, showing the staggering complexity that had arisen through successive subdivision; some fields are less than ten metres across, and a separate map shows the distribution of selected holdings, many with five to ten scattered fragments, one with fifty-five.

The minutest details of crops, rotations, agricultural operations, labour hours, crop yields, etc. are given for individual farms, although names are disguised as initials (e.g. 'R., son of L., cultivates here as a tenant but finds difficulty obtaining credit; he also plies a cart for hire'). Many fields are ploughed four to six times. We also learn of non-agricultural occupations in the village: potter, tailor, water-carrier, barber, jester, drummer, juggler, moneylender, beggar, etc.

---

[1]   At the time, part of India; some lie in modern Pakistan, others in the Punjab State of India.

Average yields for twenty different crops are given, separately for *chahi* (land irrigated from wells) and *barani* (land dependent on rainfall). Thus, converting from seers per acre, on irrigated land *kharif* (summer) maize yields averaged 1600 kg per hectare and *rabi* (winter) wheat 1200 kg per hectare, with both somewhat less than half on rainfed land. These are considerably higher than would be expected on unfertilized land today, suggesting the soils were in good heart.

There are quotations (translated from Urdu) from reports of settlement officers in the 1880s:

> This village is in good condition … The soil is quite average, in the south of the uplands a little light, in the north a grey loam with a good deal of *kallar* shown as *kalrathi*. It is liable to flooding from the *chhamb*, which on the whole does more good than harm. Population has increased since 1868.

These surveys, and comparison with the same villages at the present day, offer exceptional material for research into the historical geography and sociology of land use.

Bangladesh (East Pakistan)

> By the time I got to Dhaka in 1961, most of the districts had been covered by soil surveys. These had been carried out by chemists, without an understanding of geomorphology or any other field science. The result was 'basket-of-eggs' soil maps and pages of tabulated laboratory data (patently unchecked for reliability).

Thus writes Hugh Brammer, who had been posted by FAO to take charge of the East Pakistan section of FAO's reconnaissance soil survey of the country. On his own at first, Brammer was subsequently joined by a multinational staff from the UK, the Netherlands, Chile and Peru. The originator of FAO's system of agroclimatic classification, Amir Kassam, also served as a consultant.

Starting from scratch with staffing, Brammer's efforts at training a group of agriculture graduates field soil survey have been described in Chapter 6. When, after some initial difficulties, this was achieved, these formed the core staff of what was to become the Soil Resources Development

Institute. Also at the start of the work, Brammer had adapted the United States land capability classification scheme to the very different conditions of largely depositional tropical landforms.

Over the next ten years Brammer and his team set about producing a series of district surveys at reconnaissance level, the first in 1965. Together these contain a massive amount of data. The 1967 report for Dhaka District (formerly Dacca) may be taken for illustration. At 390 pages the largest of the District surveys, it is on bound, mimeographed foolscap. The maps show soil associations at 1:125 000, present land use at the more normal reconnaissance scale of 1:250 000, and land capability also at the latter scale. This was one of thirty-one similar district surveys completed in the period 1965-77.

Rioting connected with the independence of Bangladesh led to the evacuation of the FAO team in 1971. Reconnaissance surveys by the Soil Resources Development Institute, now working independently, continued. After a period spent writing up the FAO survey, Brammer returned to Bangladesh in 1974, initially for FAO and later as a consultant. He produced a large number of reports and articles, most of them locally published, showing how survey knowledge could be applied to a wide range of development activities, ranging from national level planning to guidance for villagers on how to produce their own agricultural development plans. A theme running through his work is the correct siting of agricultural trials and demonstration stations, so that they are representative of defined extension areas. Bangladeshi farmers operate complex crop rotations and management practices, based on an informal, farmer-to-farmer, extension system. The seven books which he produced in retirement, so that his knowledge of the country would not be lost, have been noted in Chapter 6. As he rightly observed, 'We know more about the land resources of Bangladesh than any other large tropical country.'

Sri Lanka (Ceylon)

Sri Lanka has a long Colonial history. Visited by the Portuguese in 1505 and occupied by the Dutch in 1658, it became a British colony in 1802. In 1948 it was granted dominion status, still as Ceylon, and in 1972 became the Republic of Sri Lanka, remaining within the Commonwealth.

For a country of some five million people, the Department of Agriculture in the inter-war period was small, twenty to twenty-six in headquarters, with a higher proportion of specialists compared with divisional agricultural officers. These included systematic and economic botanists, two mycologists, two entomologists, one agricultural chemist, three posts of inspector of plant pests and diseases, and the curator of the Royal Botanic Gardens.

Systematic soils knowledge began with the appointment of A.W.R. Joachim as Agricultural Chemist in 1925, whose distinguished career is outlined in Chapter 2. Working largely alone at professional level — he was granted an Assistant Chemist only in 1942 — he travelled and observed landscapes, carrying out no fewer than ninety-eight local surveys. By assembling this knowledge he was able to present a classification and a provisional soil map of Sri Lanka in 1945.

The classification combines a logical scientific structure with soil descriptions in terms easily understood by non-specialists. The basis was the concept, then current, of dividing soils into three groups:

- **Zonal soils**, with climate the determining factor. The two zonal soils represent Sri Lanka's major climatic divisions, the Wet and Dry Zones (the latter with a rainfall of 1000-1900 mm, which in other countries would hardly be called 'dry').

- **Intrazonal soils**, dominated by parent material. In Sri Lanka, these are the soils formed from limestone.

- **Azonal soils**, under which he put alluvial areas and the distinctive profiles developed under rice paddy.

Using the framework of these classes, he employed the descriptive methods usual at the time, simple but informative, such as, 'Reddish brown lateritic loams' or 'Soils similar to the black cotton soils of India'. Reflecting the compilation of the map on the basis of travel and interpolation, all parts of the country carry a shading for soil type, although there are no linear boundaries between them.

A new phase began with the Canada-Ceylon Colombo Plan project 1955-60, its first contribution being air-photograph coverage. An institutional base came with the formation of a Land Use Division, responsible for national soil survey, in 1959, whilst C.R. Panabokke, first appointed 1950, became Head of Soil Survey in 1960. The Colombo Plan project continued with soil survey of fifteen river basins in the north-west of the country, leading to a soil map of the Kelani-Aruvi area, extending north from Colombo, at 1:250 000 scale; the mapping units were catenas with local names. Frank Moorman, an experienced Dutch scientist, came as consultant under the project, and working jointly with Panabokke produced a revised classification, employed in the new national soil map of 1968. The earlier descriptive soil types were replaced by classes which linked the former US terminology with the recently developed FAO units, for example:

- Soils of the Dry Zone: Reddish brown earths (Chromic Luvisols).
- Soils of the Wet Zone: Red-yellow podzolic soils (Humic Acrisols).

By this time, pressure on land in the Wet Zone was leading to a drive to colonise the more sparsely populated Dry Zone, particularly through irrigation schemes. Sri Lanka had some remarkable early irrigation systems, developed before European influence and centred upon Maha Illuppallama, many of which had been restored by the 1950s. The major new schemes were in river basins of the Dry Zone. The largest of these, the Mahaweli Ganga, was surveyed by Hunting Survey Corporation of Canada, the report on which is a model of good studies by consultant companies.

Despite the political difficulties and insurrection which seriously hampered development, resource survey by the Sri Lankan government has continued to advance. In a 1984 FAO project, Assistance to Land Use Planning, Robert Ridgway and I helped Sri Lankan staff to start an environmental database for land evaluation, and from 1995 further cooperation with Canada, with Ranjith Mapa the Team Leader, has produced a Soil and Land Resource Information System for Land Evaluation (SRICANSOL). Sri Lanka has been well served by its local resource survey staff and institutions.

Recollections

For a piece of anecdotal material I go somewhat offline to forestry, to a research planning meeting of the International Union of Forestry Research Organizations (IUFRO) held in Peradeniya. The delegates were being shown around the famous Royal Botanic Gardens, the site where South

American rubber trees were grown prior to being planted in Malaya. The leader announced that we were now coming to the bamboo avenue. One of those present, we will call him Geoff, expressed his interest in seeing this; another commented, 'I didn't know Geoff was a bamboo scientist' to which a Malaysian Chinese responded, 'He not bamboo scientist. He donor funding bamboo research, want to see what bamboo look like.'

Christopher R. Panabokke (b. 1926)

Soil survey in Sri Lanka was never dependent on expatriates. For forty-five years, 1925-70, it was led by two outstanding officers of local origin, both of whom became Directors of Agriculture, a distinction infrequent among soil surveyors. A short biography of the first of these, A.W. R. Joachim (1898-1979) has been given in Chapter 2.

Chris Panabokke was a Sri Lankan national who belonged to the most pres-tigious group, the Sinhalese of the Kandy region. His education was a chemistry degree from the University of Ceylon, followed by a PhD in soil science at the Waite Institute, Adelaide, Australia, and soil survey experience with a leading Australian surveyor, C.G. Stephens. Joining the Department of Agriculture in 1950, he became successively Head of the Dry Zone Research Institute, Maha Illuppallama, Head of Soil Survey for ten years from 1960, Director of the Mahaweli Development Board, then in the Department of Agriculture successively Director of Research and Director General.

After retiring from government after 33 years, he spent a further period of the same length with national and international organizations. It was Panabokke who led the Sri Lanka side of the Colombo Plan cooperative project with Canada, which produced a revised soil classification and a new national map. His interests extended into agroclimatology and especially irrigation management, including a post as Irrigation Agronomist with the International Water Management Institute (IWMI), at the time recently founded and based in Sri Lanka. In 1986 he was awarded the Vidya Jyothi, Presidential Honour, the highest scientific honour of his country.

He continued working to a late age. When I contacted him in 2005 he was still, at the age of seventy-nine, engaged in consultancy. In 2010 he published an account of indigenous village irrigation systems in Sri Lanka, and in 2014, on his 88th birthday, a Sri Lankan newspaper quoted him being worried about the future of science and the scientific community in his country.

Myanmar (Burma)

From the late nineteenth century onwards Burma was administered as a Province of the Indian Empire. It became a separate colony in 1937, followed by occupation by Japan during the war. Burma became independent and left the Commonwealth in 1948; the change of name to Myanmar came in 1989.

Pioneering accounts of soils and soil-crop relationships, from 1916 onwards, have been mentioned in Chapter 2. They notably include two detailed surveys from 1932, one of the area which the outside world most associates with Burma, Mandalay.

The first national soil map at 1:2 500 000 scale was prepared by B.G. Rozanov 1959; inference played a substantial part in its composition. In 1969-73 Maurice Purnell worked for FAO on a survey for irrigation development in the Sittang Valley. He writes, 'This was some of the best field survey work I ever did, in difficult conditions and constantly under armed guard.'

11

SOUTH-EAST ASIA AND THE PACIFIC

Malaysia (Malaya, Sarawak and North Borneo)

ANYONE WHO WANTS to see soil survey put to effective practical use should look at Malaysia. For more than fifty years it has had an efficient national soil survey, carrying out both reconnaissance and detailed studies. The transition from expatriate to local staffing took place relatively early. Most importantly, a set of soil names was established which are widely recognized, and used by estates and agricultural advisers to make fertilizer and other land management recommendations. The reason for this situation is the dependence of the agricultural economy on two perennial export crops, rubber and oil palm, grown both on estates and by smallholders; research is such a relatively cheap activity that a very small export levy can maintain well-equipped and adequately staffed institutions.

Malaya became the independent Federation of Malaysia in 1957, Singapore seceding in 1965. Two territories in the island of Borneo, Sarawak and British North Borneo, became States of Malaysia in 1963, the latter as Sabah. A third territory in Borneo, Brunei, was formerly under British protection and became independent in 1983, remaining in the Common-wealth.

West Malaysia (Malaya)

Malaya had a Chemist within the Soils Division from 1921 onwards, and by the late 1940s there was a Soils Division within the country's renowned Rubber Research Institute. John Coulter was Soil Scientist 1948-57, becoming Director of Agricultural Research for the subsequent two years. He was mainly involved in soil fertility work, but also conducted surveys of peat swamp areas. Arising from experience in both the West Indies and Malaya, he wrote a particularly clear analysis of the applications of soil surveys.

The first soil surveyor to be appointed was William Panton (Bill) in 1953. By 1958 he had produced a survey of Trengganu. Panton devoted nearly the whole of a long career to Malaya, marrying into the family of a Sultanate. The chapter on soils which he wrote for Wyatt-Smith's *Manual of Malayan Silviculture* in 1963 appeared also in the revised edition of 1995, giving him a publications span of nearly forty years.

The Trengganu memoir was to become the start of a national reconnais-sance survey, organized by the component States, extending from 1958 to 1970. The authors of these include four British expatriates and five Malay-sians. The first local soil surveyor was Law Wei Min, appointed 1962, followed closely by Ignatius Wong. In the late 1960s the Soil Science Divi-sion of the Department of Agriculture was headed by the soil chemist, Ng Siew Kee, and the Soil Survey Section by Law Wei Min. A reconnaissance soil map of Malaya appeared in 1962, and, 'a very much needed revision' of this in 1968.

A national system of soil series, related to fertility and management, had been started by Coulter. It gained momentum when Ng Siew Kee attended a conference in New Zealand in 1962, passing on his enthusiasm to his colleagues. This led to a New Zealand Colombo Plan Aid project 1964-68, expanded into a Malaysia-New Zealand-Canada-USA (Peace Corps) joint effort. The leading figure in this was Michael Leamy, Soil Correlator, seconded from the New Zealand Survey. Leamy and Panton produced a *Soil Survey Manual for Malayan Conditions* (1966). This has a wider scope than its title might suggest. After setting out standardized methods of soil descrip-tion and mapping, and criteria for the definition of classes, it outlines the principal soil series of the country. Since the climate is hot and wet, and the vegetation rainforest, in all parts of the country, these series are largely based on parent material: sandstone, shale, andesite, etc. In some cases there are two on one rock type, e.g. with and without a concretionary laterite horizon, shallow or deep.

These are wider than the standard definition of soil series, but possess two advantages. Firstly, they are readily recognizable, so that estate managers can call in a soil scientist or even a general agricultural adviser and know what is on their land. Secondly, for each series a set of management recommendations has been established, for crop selection, conservation, and most importantly, fertilizer recommendations. This practical system has been applied in many hundreds of local surveys. As part of the First Malaysia Plan produced by the Economic Planning Unit, the country-wide reconnaissance soil survey was extended as a national land capability classification report (1971).

Land settlement projects: Jengka and Johor

The Malaysian Federal Land Development Authority (FELDA) embarked on a series of projects designed to take landless people from the overcrowded coastal lowlands and settle them in the forested interior. Two of these projects, Jengka and Johor Tenggara, are representative examples of land settlement schemes, for the planned use of previously unoccupied land. The capital costs were supported by loans from the World Bank. These schemes might not today have received approval, since they necessarily involved clearance of primary rainforest, Malaya's dipterocarp forest being amongst the finest in the world.

One of three survey teams ready to set off into the forest; Anthony Young centre; the *mandor* (foreman), Ramli bin Puteh, holds the kettle for making tea

The Jengka Triangle is an area of 120 000 hectares in the State of Pahang, in the centre of the country. In 1965 this was largely under forest, except for a belt of settlement along the River Pahang. The project was planned in 1965-67, undertaken jointly by the British agricultural consultants, Huntings, and the US engineering firm of Tippetts-Abbett-McCarthy-Stratton (TAMS).

Crossing the River Pahang (with limited freeboard)

Where land is covered by rainforest, air-photograph interpretation is of limited use; the trees grow taller in the valleys, presenting an almost uniform canopy. Soil survey was therefore undertaken by means of a *rentis* system, the Malay word for a cut line through the forest. This was designed on a herringbone pattern, a central north-south baseline from which equally-spaced *rentises* extended east and west at regular intervals. Soil observations were taken at regular intervals along these by a combination of pits and auger holes. The surveyors were Bill Verboom (leader), John Bennett and myself, each accompanied by a team of five Malay labourers. Field headquarters was set up in wooden Malay huts on stilts, built for the intended first group of settlers. A team would head off along a *rentis* for a week, camping in the jungle. There was no need for the usual initial stage of identifying and classifying soil types. It was agreed to adopt the existing soil series, and members of the national soil survey took the newly-arrived team to field sites where these could be seen. This situation typifies the benefits of having a national reconnaissance soil survey.

Soil survey in rain forest: neither air photographs
nor view of the landscape are available

The second major element in land resource survey was a forest inventory, a quantitative estimate of the timber that would become available from clearance for agriculture. In rainforest it is essential to have controlled exploitation, otherwise logging companies would maximize profits by creaming off only the most profitable species. Some exploitation was already taking place, the logging tracks of which provided useful access, although unfortunately the logging lorries had a high clearance, so in a Land Rover it was necessary to try and straddle these deep channels, a difficult feat in wet mud.

Logging company clearing the forest

The team were asked to convert the soil map into a land suitability classific-ation, showing suitability for oil palm, rubber, and land to be left under forest, for collection of minor forest products and a limited degree of wild-life conservation. Oil palm is the more demanding crop, whilst rubber will grow on poorer soils. Being known to have experience with the FAO system of land suitability evaluation, I was asked to prepare a first draft, drawing up criteria of slope and soil type for each intended use. On seeing the results, Verboom disagreed with some of the allocations. 'I have managed oil palm plantations. If I say land is suitable for oil palm, it is suitable, isn't it?' We were unable to employ this approach as the sole basis for land classification, but modified some of the limiting values.

The report for the Jengka project is notably compact: an Outline Master Plan, of only sixty pages; a thicker volume of text, containing all the material on soils, forestry, agriculture, and draft settlement plans; and a volume of maps. It was awarded a prize by the American Association of Consulting Engineers.

The Jengka Triangle Project went ahead, and in 1987 was the subject of a World Bank impact evaluation study. By that time, 40 000 ha of jungle had been cleared, 26 000 ha of oil palm and 14 000 ha of rubber planted, and 9200 smallholder families settled on holdings of 4 ha. The yields of oil palm and rubber were equal to those forecast in the plan, in the case of oil palm slightly higher; prices of these commodities fell slightly, but the economic rates of return were close to those projected at appraisal. 'The most important factors accounting for project success and sustainability of bene-fits have been project design, borrower support, adequate project organiza-tion and a sound settlement system.'

The project under way, based on land use planning with conservation agriculture. Foreground, oil palm on gentle slopes with rich soils; middle distance, 100% protection of sloping land with cover crop, prior to planting rubber; background, natural *Dipterocarp* forest retained on summits to reduce runoff.

Institutional effectiveness is singled out as a striking feature. 'FELDA has grown into one of the most successful land organizations in the world, combining the efficiency of the private sector with the public service of a government agency.' Another factor was the implementation of the project in three stages. Rightly, the plan was not laid out rigidly, but adaptations made during implementation. Soil erosion was avoided by correct land clearance methods, maintaining a surface cover. There was necessarily what is now regarded as a negative environmental impact, in the clearance of high-grade forest, resulting in less regular river flow and loss of wildlife. It is rewarding to those responsible to see project design recognized, and indeed, the map showing the project after settlement bears a striking resemblance to maps prepared during the survey.

A second survey, larger in scope, followed in the south of the country in 1970-71, the Johor Tenggara regional master plan. Huntings was again involved, in collaboration with three other companies, including the Overseas Development Group of the University of East Anglia, UK. This consultancy company was the brainchild of Professor Athol Mackintosh. The basis of it was that the University's School of Development Studies was staffed 33% above its establishment, members being required to spend one third of their time on externally-funded projects. This had the effect of an ongoing transfer of experience between student teaching and practical consultancy. Mackintosh himself led the Johor study.

There had been no sociologist in the Jengka team, but the University insisted that one should be employed for the Johor plan. At the time this was regarded as something of an innovation, although today it is normal (sometimes to the neglect of land resource survey). There are nine volumes to the project report. Following the regional master plan these cover natural resources (land and water), sociology, technical aspects of development (agriculture, forestry, conservation and tourism), infrastructure, and institutions. There is also a map folder, and many working papers. This report is a notable illustration of the integration of land resource studies (soil, water, and forests) with agriculture, sociology and infrastructure. Its main recommendation, however, 'The progressive clearance of 281 500 acres [114 000 hectares] of forest, and development of oil palm, rubber, dairying and other agricultural activities' would not nowadays meet with universal approval.

Recollections

There are several distinctive features to camping in rainforest. When, at some point along a *rentis*, one announces, 'We make camp here,' each labourer takes a sharp *parang* (machete, large knife) and cuts down saplings. These are trimmed into the required shapes: for the bed of each person, four Y-shaped vertical supports, two 1.8 metre sides which fit into canvas beds, and two shorter cross-pieces, bevelled at the ends to fit firmly between the sides. By rubbing the supports with tobacco or soap, the leeches are prevented from climbing. (In South America, the same effect is achieved by hammocks.) The 'tent' is a roof only, with the sides open for ventilation. The night is noisy, with a continuous racket from crickets, and during a thunderstorm one cannot hear the person in the adjacent bed. Using dead wood for cooking, it is remarkably self-sufficient.

Midway through the Jengka Triangle project, John Bennett earned the censure of his colleagues by discovering, part way through the survey, that the definitions of two of the soil series were separated according to whether a subsoil layer of laterite was less or more than 50 cm thick. This increased the labour needed to separate them, and the presence of some linear east-west boundaries between adjacent rentises surveyed by Bennett and myself suggests that one of us was being less than conscientious.

The same project took Bill Verboom, born and brought up in Indonesia, back to the jungle he had known and hunted in as a young man. On arrival he had brought two rifles, and addressed the immigration authorities in fluent Malay with an Indonesian accent. As this was a time of *konfrontasi* (confrontation) between the two countries, it was quite some time before he was allowed through. Knowing the way to do things in remote jungle country, at the start of the survey he enquired which was the most prestigious religious establishment in or near the area of survey. Informed that this was a Hindu temple, he drove to this and, for a small donation, acquired a blessing from the priest. This turned out to be a valuable precaution. Of the other team members, I went down with dengue fever, and a large tree fell across Bennett's Land Rover. Verboom remained unharmed.

East Malaysia: Sarawak (Sarawak)

If soil surveys in Malaya were well organized and effectively applied, those in Sarawak were if anything more so. With nearly the same area as Malaya but less than one sixth the population, there was great scope for new land settlement, and thus for a basis on which to plan it.

David Wall arrived in Sarawak in 1960. He had taken a geography degree at Reading, and decided that soil survey 'would admirably combine this with my farming background.' The Colonial Office accepted his application and sent him for six months' training in soil science at Aberdeen, which included field experience with the Soil Survey of Scotland, followed by three months at the International Training Centre (ITC), Netherlands, where he gained, 'an eye-opening introduction to the art and science of stereoscopic air-photograph interpretation.' It is hard to imagine a better education for a soil surveyor.

Wall was joined by Ian Scott and Jacobus Andriesse, one of a several of Dutch surveyors who contributed to British Colonial survey:

> We amicably split the country into three for working purposes, built up a strong team with eight trained field assistants, a cartographic unit, and a responsive and well-equipped laboratory under John Bailey. In the first years most of our work was to examine the potential of land for swamp rice and the key commercial crops of rubber, oil palm and coconut through reconnaissance and semi-de- tailed surveys. These bread-and-butter surveys were leavened by an occasional detailed study of a research station. Eventually we built up a national soil classification, which led to the ambition of completing a soil map of the country. In this we were supported with nearly complete coverage of RAF air photography dating from the 1950s. Between the regular surveys for future immediate agricul- tural development we fitted in reconnaissance trips of the land between, and gradually were able to fill in the complete picture, an altogether satisfying experience to conclude our time there.

The order of priority in this passage may be noted, applied surveys taking precedence. In seven years, Wall estimates that he produced 'about 40 to 50 reports,' a modestly-expressed statement which shows his considerable dedication and energy.

Most of the maps and reports listed in this book are reconnaissance surveys, but this misrepresents the balance of work in many countries. Seeking to remedy this, an example at the opposite extreme of scale can be cited. This is a 1978 report entitled, *Soil suitability for oil palm of an area of 2916.8 acres of Sarawak*. Kerasa Plantation, which sponsored this survey, knew what they wanted.

Recollections

David Wall met with one of the worst experiences a surveyor can encounter:

This happened early on in Sarawak, when I was a novice to local conditions. My wife was accompanying me, and we were following a cut line through beach ridges and swampy patches of mangrove. My field assistant pointed to a strange shape hanging from a low bough directly in our path; he said it was a hornets' nest, and we should make a dog-leg in our line. I knew better of course, and chose to walk past it. Pandemonium broke loose, as hundred of vicious insects gave chase to every one of us! With yells and screams we tried to escape in all directions, some of it through knee-deep water between mangroves. Some dived under a thick bush, others cut leafy branches to cover themselves; while I, conscious of the value of the camera I was carrying, dropped flat on my back in the muddy water with my hat over my face as protection and the camera on an island on my stomach. I had some painful bites to remind me of my stupidity, but ended a wetter, muddier but wiser Soil Surveyor.

East Malaysia: Sabah (North Borneo)

The story of North Borneo is of two surveys: the first, a one-man study, carried out with much energy and skill but inadequate support; the second, a team effort, which produced one of the most comprehensive appraisals of resource potential in the whole of the developing world.

In the 1950s North Borneo was a sparsely populated country, one of the emptiest parts being the Semporna Peninsula in the south-east. In 1951 Herbert Greene, the peripatetic Natural Resources Adviser to the Colonial Office, visited the Colonial Development Corporation estates at one of the few settlements, Tawau. The primary purpose of this visit was a problem of dieback in hemp, but Greene realised there was a potential for a range of perennial crops in this totally forested peninsula, influenced by the fact that geological survey in the 1950s had shown the presence of basic igneous rocks, an indicator of the likelihood of fertile soils.

Arising from these chance circumstances, Colonial Development and Welfare funds were allocated to a reconnaissance soil survey of the Semporna Peninsula. This was conducted by Ron Paton in 1953-59, working in isolation. After graduating in geology at Durham, Paton joined the Colonial Research Service in 1951 and was given two years' training, noteworthy in being under three most eminent soil scientists: Walter Russell at Oxford, Edward Crompton of the Soil Survey of England and Wales, and Norman Taylor in New Zealand.

Arriving in North Borneo in 1953, Paton was charged to carry out a survey of 4000 km² of forested country. This had to be conducted by the *rentis* system, and part of his first tour was spent learning enough Malay to control gangs of labourers. 'Thankfully, Dayaks came up from Sarawak on a seasonal basis and were quite marvellous; they were pleased with the arrangement, for after cutting and marking rentis lines in the field season, they used the same lines to collect *damar* gum.'

This survey is unusual in being a reconnaissance study in purpose and output, but with maps at the semi-detailed scale of 1:50 000. The results showed the existence of considerable areas suited to a range of plantation crops. In particular, there was a basaltic eruption centre with 2500 ha of soils of incomparable fertility. 'In the last two years of the work we were able to construct a block-diagram model of the whole peninsula; this not only went down well with the Governor and the visiting Duke of Edinburgh, but aroused the interest of plantation companies on the east coast.' The report is open to criticism for lack of balance; there is much technical detail, but only a two-page section on land potential. When taken to task over this, Paton replied that he would have liked to expand on the land use planning aspects, but was instructed that this was not his task.

A solo effort such as this, however dedicated, could not meet the needs of such a large country, and the Land Resources Division came to the rescue. A team of seven completed a reconnaissance soil survey of the whole country from 1970 to 1972. Air photographs at approximately 1:50 000 were preferred to larger scales, allowing a better overview of landscape contrasts and giving detail appropriate to mapping at 1:250 000. Field survey was by traverses, camping along them; the field parties included ten to fifteen assistants and two boatmen.

The report on the soils (1975) is in five volumes plus a folder of maps. Volume 1 sets out the soil types recognized, using the FAO classification. Volumes 2-5 cover four regions, each with the usual coverage of environment, soil associations, and substantial accounts of suitability for agriculture. It is exceptionally well illustrated, including colour photographs of soil profiles, and stereograms of air-photograph pairs, and the maps are very clear. The whole is a model of clear presentation.

Not content with this appraisal, the Sabah government and the Land Resources Division set up a further stage. A technical subcommittee on Land Capability Classification was set up, with forty members drawn from seven Departments: Agriculture, Forests, Lands and Surveys, Geological Survey, Drainage and Irrigation, Public Works, and the Ministry of Finance. The earlier soil survey was linked with a Canadian-financed forest inventory, and another essential for opening up new areas for settlement, a survey of road-making materials by the UK Road Research Laboratory. By assembling such a wide range of expertise and interests, a new synthesis was made, somewhat modestly termed a land capability classification. There are again four regional volumes together with a new set of maps of 1:250 000, the latter showing land best suited to mining, agriculture (high and moderate potential), productive forestry, and conservation. In the reports, as compared with the soil survey the proportions are reversed: the descriptions of soils, forests, etc. are relatively short, the appraisal of land potential more extended. The result is an outstanding example of translating survey information into opportunities for development.

Recollections

The hemp dieback problem which was the original cause of Paton's survey was to be the bane of his time in Borneo. Herbert Greene thought it was due to a trace element deficiency in the soil, collected 150 samples, and said that the man he was sending out, viz. Paton, would deal with these. However, a visiting plant pathologist suggested the hemp was suffering from 'bunchy-top', a virus spread by aphids; this view was adopted by the local plant inspector, who became known as 'Bunchy-top' Parker [name changed]. Plantation management were faced with the choice between views of the Colonial Office Adviser on Natural Resources and the local Inspector, who had no formal qualifications but local experience, and in this matter was right. The dispute simmered on for many years until the Inspector retired to Australia, as Bunchy-top Parker MBE.

Brunei

With its wealth derived from oil revenues, Brunei was in a position to commission studies of its land potential. A land capability study at 1:100 000 was completed by Huntings in 1969, a rare example of a country self-financing a consultancy.

Samoa and Fiji

These Pacific countries have in common that they were surveyed partly (Fiji) and wholly (Samoa) with the aid of the New Zealand Soil Bureau, in the person of Charles Wright. Western Samoa was under New Zealand administration from 1914, and in 1962 became the first Polynesian territory to be granted independence, dropping 'Western' from its name in 1997. Fiji was formerly a British Colony, became independent in 1970, left the Commonwealth 1987 and rejoined ten years later.

Samoa

The population of Samoa lives largely along the coastline of the two islands. Fieldwork began in May 1956 and was completed, for 3000 km^2, in December of the same year, a rate of progress which could only have been achieved in one way, by sending Charles Wright as the surveyor. The small print in the acknowledgements tells how the work was done:

> Much of the success is due to the enthusiastic help of Samoans employed as field assistants. The landscape was examined mainly by making traverses on foot, and during much of the time we were dependent upon hospitality offered by village communities.

In other words, Wright and his locally recruited and trained helpers disappeared into the interior for eight months.

Having acquired a vast amount of information, from field observation and talking to farmers, it was Wright's practice to set it all down, finding time to do so in the early hours of the morning. The result is a report which some might consider to contain an excessive amount of detail. There are maps at 1:100 000 of soils and land classification, excellent cartography making clear both the overall picture and detailed boundaries. The legend to the latter is replete with information on soil limitations, nutrient needs, and potential for agriculture, pasture production, and reafforestation. At the lower end of the spectrum, Class 4, Subclass 4c consists of, 'Soil mostly in fissures between rocks, difficult to clear for planting and to harvest'. This is not trivial: given hand labour, such soils can be fertile. I have heard a farmer say, not without scientific truth, that the rocks feed the plants.

Fiji

Fiji appointed a soil chemist, C. Harold Wright, soon after the establishment of an Agriculture Department. His accounts of fertilizer trials and soil requirements of rubber, sugar cane, cotton and other crops, published 1916-17, are among the earliest from British territories. In the 1937 a later soil chemist, W.J. Blackie, proposed a national survey, but the project was delayed by the war.

In 1946 the Legislative Council accepted that a soil reconnaissance of the whole country was a basic need for agricultural development. The government of Fiji sought assistance from New Zealand, which drew attention to the prior need for air photography, and as soon as some progress had been made with this, sent two surveyors in 1952. Meanwhile Ian Twyford from England had been appointed as Soil Survey Chemist, to collaborate with the field team. After further breaks through staffing problems, Charles Wright arrived in 1955. With his immense energy he accomplished the greater part of the field survey, assisted by Twyford and a locally trained field assistant, Voluma Tora. After completing the two main islands, most of the smaller islands were surveyed by Twyford and Tora.

The *Soil Resources of the Fiji Islands* (1965) is a massive account, accompanied by a box of maps. These consist of a soil map and a land classification map, each in eight sheets plus a legend. Part I of the text covers standard material: environment, soils, and past and present land use. Part II, 'The soil resources and land use of individual islands', occupies 270 pages of closely-printed text written in Wright's inimitable style. For each island, there is a detailed description of the soils locally represented, with exact areas covered, followed by equally thorough accounts of their fertility and other management properties, hazards such as erosion, present use, and potential for improved use and management. Some indication of the nature of this material can be gained by a highly abbreviated extract from the account of one of the smaller islands, Kadavu:

The Fijians of Kadavu are farmers of repute. In the old days they were famed for skilful terracing and hillside irrigation projects ... Owing to regular and totally unnecessary burning, the former forest cover on the leeward (northern) side of the island has been replaced by reed grass ... Bananas and copra are widely grown. The most suitable permanent crops are coffee, cocoa and citrus, although these will be limited by the low fertility of many of the soils.

This is followed by fifteen pages covering each of the fifty-two soil types and phases mapped on the island: characteristics, land use, and potential for development. Detailed information of this kind provides a source of information for agricultural development and extension in each specific area. The very length and density of the text leads to the only possible criticism, that it is difficult to obtain an overview of the main features, to see the wood for the trees.

A later study, the *Fiji Forest Inventory* (1973), one of six forest inventories made by the Land Resources Division between 1971-74. With maps at 1:50 000, this gives areas covered by forest and quantitative information on timber resources. An inventory of this kind is an essential basis for planning forest conservation and utilization, although in practice, prevention of unauthorized clearance and illegal logging is difficult in the social and institutional set-up of most developing countries.

Recollections

Some years ago in a review of the US Soil Taxonomy I illustrated its bulk by the fact that it weighed 2.25 kg. It is said that compilation of this involved more than 100 000 man-years. The two parts of the Fiji memoir written by Wright and Twyford amount to 2.5 kg.

When Twyford was doing fieldwork in a remote forested area of Viti Levu, staying in a Fijian village, there was an evening entertainment of songs and dances (*meke*). He realised that one *meke* was about future happenings, concerning things which the Fijians could not possibly know. They told him that time was a continuum, the future and the past were one. The events they foretold came to pass about a year later. He found out afterwards that such happenings were not unusual.

Papua New Guinea

In Papua New Guinea there is a rich indigenous knowledge of soils, in terms which scientists would call their fertility, workability, and suitability for crops. This is in part due to dependence on root crops, and associated diversity of crop varieties and cropping practices. They are aware that falls of volcanic ash bring added nutrients. Geoffrey Humphreys, following soil studies in 1979-1981, reports, 'The women, especially the older women, appear to employ a more detailed terminology that is largely beyond the comprehension of males.' A feature of the country which is not wholly explained is that, despite the high rainfall and many steep slopes, soil erosion is not such a serious problem as might be expected.

Soil, land evaluation, and crop suitability studies have been many and varied. Two institutions were involved, the Soil Survey Section of the Department of Agriculture, established in 1948 under the direction of G.K. Graham, and CSIRO Australia.

The Government Soil Survey conducted local studies for specific development purposes such as tea and coffee estates, or smallholder oil palm and rubber schemes. As usual for surveys of this type, many of the proposed developments were implemented. An example is a survey of the Markham Valley, in which special attention is paid to the soil requirements of sugar cane, and was used in the siting and layout of the country's first sugar cane plantation at Ramu.

Land system survey, the well-established technique of reconnaissance mapping which had been carried out in the northern and interior territories of Australia, was applied to the Australian mandated territory of Papua New Guinea from 1953 onward. The Division of Land Research and Regional Surveys, CSIRO Australia, produced a series of fifteen regional surveys, mostly at scales of 1:250 000 or 1:500 000. They covered a major part of the populated areas of the country, together with some empty land thought to have a potential for settlement. Excellent continuity was achieved through two scientists of Dutch origin, Herman Haantjens working from 1952-1977 (Project Leader 1954-1972), and Pieter Bleeker from 1965-1982. This extended project was notable for making advances in the application of land evaluation to land system surveys. The results were brought together in summary reports in the early 1970s, leading to a map of agricultural land use potential. Collaboration on land resource evaluation and development between the UK, Australia, and Papua New Guinea continued after independence in 1975.

Pacific islands

The United Nations Conference on Environment and Development (UNCED), 1992, in its report *Agenda 21* (agenda for the twenty-first century) recognized island states as a sensitive environment with special problems, notably extreme shortage of land, water scarcity, and often dependence on one export crop. Because of these problems, and also the fact that their small size helps to accomplish more from a limited amount of investment, they often receive favourable treatment from aid agencies.

The Land Resources Division undertook a comprehensive survey of the Solomon Islands; the report (1974) is in eight volumes, together with twenty supplementary reports giving soil descriptions, rainfall data and other supporting detail. The former joint Anglo-French administration of the New Hebrides (after independence, Vanuatu) led to the unusual situation of surveys by both British and French organizations: a forest inventory (1971) by the Land Resources Division, and a map of agronomic potential by the French equivalent, the Office de la Recherche Scientifique et Technique Outre-Mer (ORSTOM). The text is in two languages: take the memoir with its English title, turn it upside-down and you get the French version. The New Zealand Soil Bureau took responsibility for surveys of the overseas territories formerly under its administration, including the Cook Islands, Tonga, and three small and very isolated islands surveyed by Charles Wright.

The palm for the smallest and most remote territory to be surveyed must go to Pitcairn Island. In 1955 Ian Twyford was sent to conduct a geological, soil and vegetation survey, and to comment on whether it could support its people in the event that other sources of support, such as the issue of commemorative stamps, should fail. The island covers 47 km^2 and had a population of 58. He found 'incredibly fertile soils' but most of its former magnificent trees had been cut down to carve into curios. With no replanting and abundant goats the result was severe soil erosion. In the mid-nineteenth century the whole population had been converted to Seventh Day Adventism, and had become very fundamentalist. They believed the

Messiah would soon return and establish Eden in faithful communities, so there was no point in checking erosion and planting long-term species. Twyford persuaded the Pastor to deliver a sermon to the effect that the people should first do their part. That afternoon the whole community planted tree saplings, singing a hymn and saying a prayer at each. Twyford's report, of which very few copies exist, has six maps at a scale of 1:11 520, the largest scale for complete national soil survey coverage.

Other island territories

Early in its history, in 1968, the Land Resources Division completed two surveys of islands in the Indian Ocean. A soil survey of the major island group, the Seychelles, was carried out by the C.J. Piggott, and a special-purpose survey was completed, charmingly entitled, *An Investigation of the Coconut Growing Potential of Christmas Island*. A soil map of the Falkland Islands, sponsored by the British Admiralty, was produced by S. Brenna in 1946, the first post-war survey. Malta, as a developed country, falls outside our terms of reference, but exceptionally the Colonial Pool of Soil Surveyors, in the person of David Lang, completed a soil survey in 1960. In the same year a survey of Hong Kong, under British administration for a century ending 1997, was carried out by Charles Grant.

12

MAPS, ROCKS, CLIMATE, PLANTS, AND LAND USE

Maps: topographic survey

NEARLY ALL GOVERNMENT ACTIVITIES rely on maps. The earliest requirement was for cadastral purposes, or surveys of land ownership and rights of use, including a host of purposes such as local administrative districts, mineral rights, forest reserves, conservation areas, and tribal grazing lands. In the era of development, settlement schemes and other projects needed a basis of reliable maps. Departments of Survey were among the first specialist services to be set up in the Colonial Service.

The Colonial Survey Departments arose out of Britain's Ordnance Survey. In 1858 the Director General received an instruction from the War Office:

> Lord Panmure is desirous that you direct an early attention to the subject of Colonial Surveys ... reporting whether it may not be possible to establish a system under which your department ... may assist in their systematic prosecution.

This far-sighted suggestion was slow to be implemented. Interest was stimulated by the lack of maps during the Second Boer War in South Africa (1899-1902). In 1928 Brigadier Winterbotham, a future director general of the Ordnance Survey, made a tour of the Empire to assess the various survey departments, and reported on their limited progress and lack of coordination. In the 1930s Brigadier Martin Hotine pressed strongly and repeatedly for a central organization, early planning for which took place during the Second World War.

In 1946, supported by funds from the Colonial Development and Welfare Act, the Directorate of Colonial (Geodetic and Topographic) Surveys was established, with Hotine as Director. It was moved to Tolworth, Surrey, in 1951, and was later renamed the Directorate of Overseas Surveys. Hotine continued at the helm until 1965. A number of priority calls on its services soon arose: for the Volta Dam project in Ghana and the Kariba Dam on the Zambezi, the Central African Rail Link from the copperbelt of Zambia to the coast, the Groundnuts Scheme in Tanzania, and for military purposes during the Mau-Mau State of Emergency in Kenya and the Communist insurrection in Malaya. Over and above these, systematic mapping of the colonies was being undertaken.

All topographic surveyors know that it is no use starting with one local area and extending this progressively; cumulative errors build up, and adjacent surveys do not meet. A principle which Hotine initially had to fight for, administratively and in budget terms, was the need for a geodetic frame-work, based on triangulation and extending between countries. Before the war, detail in between triangulation points had been filled in by plane-t-abling, an activity that was almost sacrosanct to surveyors at the time. However, its continued use would have led to progress across the colonies being far too slow to meet demands, and made staffing requirements prohibitively expensive. Electronic distance measuring equipment became available in 1957.

The experience of air photography for reconnaissance purposes during the Second World War came to the rescue. From 1946-1953 the RAF took extensive air photographs in all parts of Africa and in the Far East. Besides the intended use for mapping this now forms an historical record of land use. To make accurate maps from air photographs it is not sufficient simply to join them up, as scale varies from the centre to the margins of each photograph. In the early days, before sophisticated machines were devised, mapping called for the remarkable procedure of the controlled print laydown: slits were cut radially from the centres and a large floor was covered with photographs joined by pins, the whole skeletal structure fitted to triangulation points.

Print laydowns formed the basis for the initial mapping of detail. Surveyors would then go out into the field, check and amplify the detail, and obtain names by enquiry. It is said that an occasional stream can be found labelled with the local word for toilet, and I have seen on maps of Malawi villages called Sindidziwa and Kaya, both of which mean, 'I don't know'.

Centralization of activity through the Directorate meant that decisions could be taken on standardization. For the map projection, the Transverse Mercator was adopted. For the basic scale, they fortunately did not persist with the British 'one inch to the mile' but adopted 1:50 000 as a basic scale, or 1:125 000 in sparsely settled areas. The first-edition maps are fairly simple sheets in black and white, many consisting largely of rivers, roads and tracks, and a widespread scattering of a tree symbol representing 'bush'. Of key importance for later use in soil survey was the fact that the principal points (centres) of air photographs are marked and numbered on the maps.

Dedicated field surveyors were called for, skilled but 'willing to live an itin-erant life in arduous conditions'. As usual with field activities, they travelled with technical assistants and porters. A large sunshade umbrella was carried, which was 'for the instrument, not the surveyor'. Hotine ruled the Direct-orate more autocratically than would nowadays be possible, and would not allow wives to accompany surveyors to their countries of posting. He held that the job of surveyors was to work in the bush for as much of the time as possible, and the presence of a wife would distract them from that objective. This no-wife rule was relaxed in 1956, and once they were allowed to do so some wives went on safari, earning their keep as recorders of field notes or catering managers. Field surveyors were considerably outnumbered by headquarters staff: mathematicians to perform the calculations involved in triangulation (they were known at the time as 'computers'), photogram-metrists, and the most numerous, cartographic draughtsmen.

The origin of the Land Resources Division as an offspring from the Direct-orate of Overseas Surveys has been described in Chapter 3. In 1968 the Division separated from the Directorate. In 1964 the Directorate supplied assistance to sixty-one countries, the highest recorded for any year. In-ser-vice training of overseas students became an important function. Following government economies in the 1980s, the Directorate was considerably reduced in staffing, and returned to the parent from which it had sprung, becoming a Division of the Ordnance Survey.

Rocks: geological survey

Geological maps are needed for a wide range of constructional purposes — roads, bridges, foundations, harbours, aerodromes — and where the materials for these can be found. They supply guidance in the otherwise chancy business of siting of boreholes for water. Then there is the aspect of most interest to governments, whether mineral deposits can be found which will supply the golden egg to support the economy.

Before the Second World War, the larger colonies independently appointed government geologists. Among the earliest was E.J. Wayland, appointed to Uganda in 1919, and Frank Dixey, appointed to Malawi in 1922. Dixey was to make fundamental contributions to knowledge of the African Rift Valley. In the absence of adequate topographic maps, the geologists made their own. Mineral investigations took priority, and without air photographs, rates of progress in mapping were slow. Thus by 1947, Kenya had mapped only 45 000 km^2 or less than 10% of the country, Uganda less than 5%.

The speed of mapping was transformed by three developments which followed the war: maps, air photographs, and a centralised institution. Topographic mapping by the Directorate of Overseas Surveys removed much effort from the geologists. Air photography by the RAF meant that many geological formations could be seen through their expression in landforms, especially in dry regions.

In this as in so many other fields of Colonial endeavour, a rapid expansion of activities took place remarkably soon after the war. The key date was 1 January 1947. This saw the foundation of the Directorate of Colonial (later Overseas) Geological Surveys, headed by Frank Dixey. Headquarters were at the Imperial Institute, London, which provided a centralized administration, coordinated staff postings, and laboratory facilities. A photogeological section was formed in 1949, and held annual courses. The initial staff of fifty officers overseas increased to 212 by 1956, largely financed by Colonial Development and Welfare funds.

Following these developments, rates of basic geological mapping rapidly increased. Thus in ten years, Kenya had covered three times the total pre-war area. Mineral investigations continued to be the major stimulus.

Systematic geological mapping had become well advanced in most territories by their dates of independence.

In 1965, the Directorate was amalgamated with survey in Britain to become a branch of the Institute of Geological Sciences (later renamed the British Geological Survey). Since that time the branch has been active in over forty countries, mapping 1.7 million km^2, or seven times the area of the UK. In addition, scientific and staffing assistance has been given through Technical Cooperation Programmes, most widely to ex-colonies but extending also to countries outside the Commonwealth.

Landforms: geomorphology

Geomorphology, the study of landforms, occupies a somewhat anomalous position in the science of the environment. Every soil survey is in fact a soil-landform survey. Because they can be seen, landforms are central to soil mapping, both for air-photograph interpretation and in the field. Landforms also play a key role in photogeological mapping and are essential to hydrological investigations, yet there were never any Colonial Officers appointed as geomorphologists. The mapping of landforms alone is hardly ever carried out. A method is available, called morphological mapping, devised at Sheffield University and practised for a time in Eastern Europe, but it was never widely adopted.

On the other hand, geomorphology plays a central role in soil survey, both at the stage of air-photograph interpretation and in the field. The land systems method rests on identification of regions jointly from landforms and vegetation, with other natural resources filled in subsequently. The first task of a soil surveyor is to determine the local soil-landform relationships. In the 1950s, geomorphology acquired an unduly prominent place in UK geography departments, at the expense of other branches of the environment. Observing the large number of postgraduate geomorphologists, some of them turned to soil survey for a career. The Land Resources Division was unusual in having posts for geomorphologists as such, linked with soil survey in a soils/geomorphology section. Soil survey was well served by the geomorphological training of many of its practitioners.

Climate: agroclimatology

When FAO, not content to rest on its laurels after completing a soil map of the world, turned to the assessment of agricultural potential, a striking truth emerged.

On a world scale, climate was found to be considerably more important than soils in determining crop production potential. In retrospect, this should have been obvious. Everyone is aware of the major climatic and vegetation zones of the tropics: humid or rainforest, subhumid or savanna, semi-arid, and desert. The crops and agricultural potential of each are highly distinct, soil types forming subdivisions within these climatic belts. Yet because of the time-honoured status of soil science, justifiable within the boundaries of many countries, FAO had devoted much less attention to climatic analysis.

Most colonies had Departments of Meteorology, initially with the task of collecting data. The white Stephenson Screen became a common sight at district headquarters. Rainfall is highly variable over short distances, and in this respect the network of recording stations was supplemented by an army of voluntary observers, since local British expatriates liked nothing better than to have a rain gauge in the garden. (One hopes the story that the gardener felt that bwana would be disappointed at having nothing to measure and topped it up, did not happen often.) Clearly what matters is not rainfall alone but its relation with evaporation and plant transpiration. It was possible to calculate potential evapotranspiration from a complex formula devised at Rothamsted by Howard Penman, but the validity of this method for the tropics was questioned. Evaporation pans, open circular tanks one metre in diameter, were added to some meteorological stations. By the early post-war years it was generally possible to get hold of maps of temperature, rainfall, and less commonly, the ratio between rainfall and evapotranspiration.

Agroclimatology, the application of climatic data to agriculture, was developed mostly at universities and research stations, without regular transfer of methods from one country to another. An important question for farmers is when to start planting at the beginning of the rains. This can be a chancy matter, with dry spells following early showers. In Uganda, for example, optimum planting dates for each crop were calculated from weekly rainfall data. From Northern Nigeria comes a story of how farmers see things differently from scientists. On a World Bank project in Sokoto State, agriculturalists found that the yield of cotton was higher if the crop was planted early, at the very start of the rains, and urged farmers to do this. Because this advice was commonly ignored, the Bank sent a consultant to talk to the farmers. He found an intricate system. The food crop, sorghum, was more important to farmers than the cash crop, cotton. At the first fall of rain, they planted sorghum on only part of their land, and widely spaced. If a dry spell followed, they had not lost much of their seed. If the rains set in, they infilled this area and planted sorghum on the remaining land. Only once the food crop was firmly established was cotton planted.

What was to become a major advance in agroclimatology started at the Institute for Agricultural Research, Ahmadu Bello University, Northern Nigeria, when Amir Kassam was appointed Research Fellow. The Director of the Institute, Matthew Dagg, set up a wide-ranging programme of agricultural ecology, with some emphasis, in this semi-arid region, on water-balance studies. Kassam, born in Zanzibar, had taken a PhD in crop-water relations of field beans. Working jointly with a soil scientist, Jan Kowal (Poland) they produced a classic study, *Agricultural Ecology of Savanna* (1978), in which the respective roles of climate and soils were treated in a balanced manner. On the strength of it, Kassam was invited to join the International Crops Research Institute for the Semi-Arid Tropics (ICRISAT), Hyderabad, India.

This experience led to Kassam being taken on as consultant in agroclimatology for the FAO Agro-Ecological Zones project. After starting with regions based on their temperature regime, a system was needed for determining zones of water availability applicable at a world scale. These would form the first stage in crop suitability, with soil distributions subsequently overlaid. The basis adopted was the growing period, the dates between which moisture is available in the soil. This is determined by the balance between rainfall and potential evapotranspiration, allowing for soil moisture storage after the end of the rains. The boundaries for equal-length periods of moisture availability (e.g. 90-150 days) formed a standardized basis for agro-ecological zones. This was combined with the moisture requirements of crops to give crop suitability regions. There had been earlier systems for mapping agroclimatic zones. An example is the set of moisture zones, I-VIII, devised in Kenya, which is valid within the country and remained in use for many years. Internationally, however, the FAO agro-ecological zones method, essentially due to Kassam, has supplied a widely-adopted basis for agroclimatology.

An aspect which combines water resources, landforms and soils is catchment studies, the effect of changes in land use, particularly forest clearance, on water flow regimes. This was developed by Charles Pereira (Perry). Whilst taking a degree in mathematics and physics, Pereira visited Rothamsted Experimental Station and met Howard Penman, who gave him a draft of his forthcoming paper on potential evapotranspiration, inspiring Pereira to work in soil physics. In 1946 Pereira joined the Scott Agricultural Laboratories, Kenya, assigned to coffee research and working on water balances. He moved to the East African Agriculture and Forestry Research Organization (EAAFRO), Muguga, near Nairobi, subsequently becoming its Deputy Director. After the Mau Mau Emergency, Pereira started catchment studies, in collaboration with the UK Institute of Hydrology. He became the leading authority on the hydrological consequences of land use changes, summarizing his work in *Land Use and Water Resources* (1973). His later career was in the UK, as Director of the East Malling Research Station, in Kent, for fruit trees, and Chief Scientist to the Minister of Agriculture.

Both academically and in terms of honours, Pereira became one of the most highly ranked of all land resource scientists: DSc, FRS, and in 1977 a knighthood. In retirement he took a leading role as President of the newly-established Tropical Agriculture Association. He also became convinced that the advances which scientific research could bring to the tropics were being counteracted by population increase, applying this belief by devoting time and energy to support Marie Stopes International.

Plants: ecological survey, pastures, and forests

Ecological survey and vegetation mapping

Ecological survey got off to a flying start with the pre-war studies in Zambia by Colin Trapnell. The conditions were favourable to his approach: large areas of empty or sparsely-settled land, with the natural or semi-natural vegetation. Although he was subsequently given the opportunity to train a small group in his methods, the approach never took off. The reason is clear: in post-war conditions land was increasingly occupied, with the natural vegetation reduced to small patches and hedgerows. The ecological approach became transformed into the use of indicator plants in soil survey.

There were some post-war ecological surveys. George Jackson's work in Malawi has been noted in Chapter 8. Among the first surveys by the fledgling Land Resources Division (then the Forestry and Land Use Section of the Directorate of Overseas Surveys) was an ecological study of the coastal swamps of Sierra Leone as a basis for expansion of rice-growing areas, conducted in 1958 by Martin Brunt. Brunt's subsequent surveys were varied in nature and objectives, but often had an ecological element.

Rowland Moss, working in South-west Nigeria, developed a method which he called the study of contemporary functional relationships. He began by arguing that the land systems approach, being based on geomorphology, was inevitably static and therefore unsuited to the dynamic nature of relations between environment and land use. He then put forward the proposition that in densely-settled areas the distinction between vegetation and land use is not meaningful. Every area of land consists of a geo-ecosystem, many elements of which are in a state of change. Soil organic matter, for example, varies between cropping and fallow periods, and may be in a state of long-term decline. Perennial crops may be ageing, attacked by disease, or in decline as a consequence of a fall in market prices. In the region of study, these factors had led to progressive replacement of cocoa by kola. Moss presents a convincing case for the area on which his conclusions were based. What is not clear, however, is how this approach can be translated into practical survey procedures.

There is a substantial science of vegetation mapping as practised by biologists, with different methods emphasizing vegetation communities, physiognomy and structure, and floristic provinces or plant geography. Its applications lie mainly in national parks and other conservation areas. Outside such areas, it is not appropriate as a basis for land resource survey for the reason already noted, that human impact on vegetation is so great. The best example is a vegetation survey of 100 000 km² in South-west Kenya, started by Martin Brunt in 1959 (published 1986), intended as a basis for expansion of cash crops, based on ecological zonation.

An unusual case of vegetation mapping was the Nigerian Radar project (NIRAD). This employed the technique of Sideways Looking Airborne

Radar (SLAR), which has the property of being able to 'see' through cloud. In 1976-78 the whole of Nigeria, 924 000 km², was mapped by Huntings, led by David Parry. This served as a demonstration of the potential of SLAR to map forest resources in permanently cloud-covered areas, although the usefulness of the survey is questionable.

Ecological survey for purposes of wildlife management and conservation can be illustrated by the Luangwa Valley, Zambia. A Game Department was formed in Zambia in 1933, headed for over twenty years by a former District Officer, T.G.L. Vaughan-Jones. A Luangwa Valley Game Reserve was declared in 1938, later divided into four National Parks, covering 15 630 km², surrounded by Game Management Areas. An early focus on safari hunting and meat production evolved into tourism and wildlife conservation. Over twenty-five years, 1965-1990, a series of studies of vegetation and wildlife habitats were conducted, some undertaken as part of an FAO conservation and development project. These were paralleled by counting the animals, in a series of large mammal surveys.

A leading role was taken by William Astle (1932-2006). In 1957 Astle had joined the government as Pasture Research Officer, but soon transferred to become Biologist and Chief Wildlife Officer. This led to a lifelong interest in the vegetation and conservation of the Luangwa Valley. In 1989 he produced a *Landscape and Vegetation Map* of the area, which became a resource for scientists and serious tourists alike. He returned to Luangwa after retirement, and wrote a history of its wildlife conservation and management, from the establishment of the British South Africa Company (1889) to the present. His obituary records, 'Bill's work was not a job; rather it was a vocation.'

Pasture resource survey

Pasture resource is obviously highly significant in semi-arid areas where extensive grazing is found. Mention has been made of surveys by Verboom, Brunt and others in Zambia, the studies of nomadic pastoral systems by de Leeuw, and the Botswana Tribal Grazing Lands project. An outstanding study was made by D.J. Pratt and M.D. Gwynne, *Rangeland Management and Ecology in East Africa* (1977). In general, however, range management specialists do not show any enthusiasm for surveys as such. The national statistics for permanent pasture which used to be produced were highly unreliable estimates, often remaining unchanged for long periods, and FAO later ceased to list them.

A central concept, although a disputed one, is that of pasture carrying capacity, the density of livestock which can be supported without degradation. The management system, e.g. annual and seasonal movements, inputs if any, needs to be specified, and even then the high year-to-year variability of rainfall in semi-arid regions leads to difficulties. When FAO was converting the *Framework for Land Evaluation* into sets of more detailed guidelines for each type of land use, extensive grazing presented the greatest difficulty. The fact that the direct land users, the animals, can move from one place to another raised problems for evaluation of the land. The last of its sets of guidelines, *Land Evaluation for Extensive Grazing* (1991), is as good a solution as any.

Forest inventory

A good-natured rivalry took place between Colonial Departments of Forestry and Agriculture, even where administrators had made them parts of a combined unit. Forest Reserves were gazetted, mostly sited on plateaux, hills and the poorer soils, often consisting of plantations for timber production from which the local people were forbidden access. Forestry Departments staunchly resisted attempts to convert reserves to agricultural land, both officially by Agriculture Departments and through illegal incursion. With the rising population pressures of recent years the protection of forest reserves, through patrols by rangers, has become more difficult.

For international statistics on forest areas, and thus rates of clearance, reliance formerly had to be made on reports from national departments. These lacked adequate facilities, and would often close an eye to incursion. The situation has been transformed by satellite imagery, allowing monitoring of forest clearance on a world scale. Successive FAO surveys have revealed that the annual loss of forest area in the tropics is about 15 million hectares or 1% of what remains. In addition to reduction of area, the remaining forest may suffer from fragmentation, limiting its value as a wildlife habitat. The Tropical Forest Action Plan and other attempts to stem this decline have not been notably successful.

Forest inventory is an assessment of forest resources, primarily the marketable timber, as a future stream of output. After surveying the area of forest, the ages and girths of standing trees are measured, and an estimate made of their rates of growth, and thus when they will reach an age suitable for felling. Softwood plantations in the tropics may reach maturity in some twenty to thirty years, rainforest or teak plantations within it in much longer periods. In Malaya and Burma, a felling and regrowth cycle of some eighty years was formerly practised, a notable example of planning for sustainable land use. The former emphasis on wood production has given place to a concern with a wider range of forest products and services, including water catchment protection and conservation of biological resources. The Forestry Section of the Land Resources Division conducted inventories in the New Hebrides, Fiji, Belize, and the Bahamas. FAO's *Land Evaluation for Forestry* (1984) successfully applied the approach that had been initially developed for agricultural land use.

One method employed in forest inventory is of wider interest. The *site index* is a measure of the quality of land carrying plantation forest. It is assessed by measuring the density, height and girth of trees across a sample area of known age since planting, and thus assessing growth rates. These are grouped into site index classes, which can be compared within and between plantations, indicating which areas are best suited to production forestry, and which are marginal. Thus the quality of land for this purpose is assessed purely from performance data without, in the first instance, soil observations. By means of subsequent soil survey on the same sample areas, the soil properties which favour tree growth — often depth and ease of rooting — can be found. This approach could be adopted in agriculture. A uniformly managed crop of, say, maize, is planted, sample blocks are harvested to obtain yield, after which soil observations are taken. This procedure reverses the usual order of activities in soil survey, mapping soil types and then searching for performance data. This has been done occasionally on experiment stations, but could be adapted to project surveys.

Land use

It has been noted in Chapter 3 how the Land Resources Division, which was to become the UK's major institution, arose out of Dudley Stamp's valiant but unavailing attempt at a World Land Use Survey. Shortly before the transition between these institutions, Martin Brunt made a land use survey of Mauritius, one of the few completed studies by the World Survey.

Shortly after, a vegetation and land use survey of the Gambia was among the first projects of what was to become the Land Resources Division. The nearest to a national inventory was the study of Malawi, regrettably unpublished, noted in Chapter 8. Apart from these, however, there were no further land use studies by the Division. Within Colonial Departments of Agriculture, there was no demand for surveys of land use.

Many soil surveys include a map of present land use, but these show the types of use: farming systems, forest reserves, etc. They do not attempt quantitative estimates of areas cultivated. This neglect is strange. One might have thought that the present land use was a second plank, to be set against land resources in the assessment of development potential. Yet land use statistics for developing countries are unreliable, sometimes ludicrously so. Large revisions are sometimes made retrospectively in the light of new data. Thus the forest area of Zimbabwe was given as 19 million hectares in the 1992 edition of *FAO Production Yearbook*, but 8.8 million in 1993. For the same dates, the cropland area of Kenya was changed from 2.45 million to 4.52 million hectares. In the major study, *World Agriculture: towards 2010*, FAO could not believe the raw data shown by its own statistics, and 'adjusted' the cropland area for sub-Saharan Africa from 140 to 212 million hectares.

Is there really spare land?

An unfortunate consequence of this unreliability of data is the supposition that there exist large tracts of 'spare land', areas which are cultivable but not presently cultivated. The source of this premise is a series of studies conducted by FAO in which two sets of data, originating independently of each other, were compared. Cultivable land was assessed from the agro-ecological zones survey, based on a climatic inventory combined with the digitised version of the soil map of the world. Cultivated land was taken from FAO's own statistics, produced by a different section and based largely on data supplied by national governments. These estimates show, for developing countries as a whole, a cultivable area of 2500 million hectares compared with the presently cultivated area of 900 million hectares. After subtracting 12% for protected land (nature reserves, etc.) and 3% for settlements, the difference is reduced to 760 million hectares, called the 'land balance'. The existence of this supposed land balance provides a comforting thought: if all else fails, recourse can be made to the age-old solution of taking more land into cultivation.

Reasons to doubt these estimates arise from what can be seen by anyone who travels and uses their eyes. If there is so much spare land, why has cultivation been extended onto steep slopes, and into semi-arid zones liable to frequent crop failure? Why has farm size in many countries fallen below one hectare, and why are infertile soils which need rest periods cropped continuously? Above all, why do 800 million people suffer from constant hunger and recurrent famines?

There are three reasons for the discrepancy between the official estimates and field observation. The first is overestimation of cultivable land. Detailed soil maps show numerous inclusions of uncultivable land — rock outcrops, scarps, small water bodies — within areas mapped with fertile soils; these inclusions get lost when the maps are reduced to small scales. The second reason is underestimation of land presently cultivated. Governments do not conduct land use surveys, they do not want to draw attention to illegal incursions onto protected land, so they frequently report the same cultivated areas for ten or more successive years. The third, and probably the largest, source of error lies in making insufficient allowance for demands on land for purposes other than cultivation. Large parts of rural areas are producing livestock from pastures and timber from forests, both essential for the welfare of the people. There are other uses such as strategic water catchments, or homelands for indigenous people. Two thirds of the supposed land balance is found in fifteen countries, many of which have large regions under rainforest, clearance of which is strongly opposed by public opinion.

I conducted a sustained campaign in print casting doubt on the FAO data, and estimating the adjustments that should be made. If these arguments are accepted, they show that in countries with a 'land balance' of less than 50% there may be little or no remaining land that can be sustainably cultivated. In countries with larger balances, the true amount of spare land may be less than half the official figures. Colleagues working in the field confirm the overall situation, that in many regions it would be impossible or unwise to take further land into cultivation. Data show that since 1995, the arable area in developing countries has remained static at about 860 million hectares, whilst the (slightly less unreliable) statistics for cereal cultivation show a fall.

In the more recent review, *World Agriculture: towards 2015/2030*, FAO have recognized that their estimates have been questioned, 'in some quarters'.

There is no need to depend on speculation of this kind. A straightforward procedure for determining the land balance of a country would be to take the official estimate, visit the country and ask to be shown which regions may have spare land, then visit these and make sample surveys. Satellite imagery is particularly well suited to land use mapping, and could extend the results of field study.

13

FROM SOIL CONSERVATION TO CONSERVATION AGRICULTURE

IN PARALLEL WITH AGRICULTURAL extension services, most countries possessed a soil conservation service. This separation of responsibilities was unfortunate, since in the eyes of the farmer the extension staff were giving advice which directly led to higher production, whilst the conservation officers seemed to be telling them to put in a great deal of additional labour with no benefit in the short term. This division was deeply entrenched, partly originating from the education of conservation staff in agricultural engineering. The modern approach of treating soil conservation as conservation agriculture is bringing the two aspects together.

The perceived threat of soil erosion was one of the major reasons for the drive to find out more about soils and land resources through surveys. In addition, survey, conservation and extension come together in land use planning. It would be possible to write a history of pre- and post-independence soil conservation services in as much detail as the present account of soil survey. This chapter will be limited to an outline of main trends with some regional examples.

Awareness of erosion

The effects of forest clearance have long been observed. Even in classical times one finds a statement to the effect, 'Do not suppose that the barren hillsides which we see [in the Mediterranean] were always like that; once they were covered with rich forests and fertile soils, and streams which are now dry flowed freely.' Cases of early environmental concern in the West Indies have been noted in Chapter 9. World attention was drawn to erosion by the 'dust bowl' in the United States during the 1930s, which led to the establishment of systematic methods of research.

It is by no means the case, however, that awareness of erosion diffused only from the USA. In the 1930s and 40s there were many reports of concern over erosion in Commonwealth territories. Among the earliest was by Arthur Hornby in Malawi, setting out the effects of erosion in reducing the capacity of the land to support the people, and measures which should be taken to combat it. Writing in 1924 and 1934, he not only presented his reports internally in bulletins of the Department of Agriculture but also publicised them in the local newspaper, the *Nyasaland Times*. Hornby in turn quotes evidence given to a Land Commission in 1920 by the Reverend Dr Laws, who had been resident in Malawi for many decades. Laws stated that the northern areas 'were well wooded only fifty years ago and streams flowed all the year round.' Hornby considered that:

> Owing to the denudation of woodland areas … the country can be said now to be incapable of supporting one half of the population it did one hundred years ago. It will not be many years at the present rate before the limit of possible food production will be reached.

Frank Stockdale published a summary account, *Soil Erosion in the Colonial Empire*, in 1937, and the following year the Imperial Bureau of Soil Science, Rothamsted, issued a Technical Communication, *Erosion and Soil Conservation*. Writing in Lord Hailey's *An African Survey* (1938), Elspeth Huxley assembled reports of other African territories, including descriptions of erosion in Kenya, Uganda and Tanzania. In Lesotho, gullying was said to have destroyed 10% of the arable land. In 1935, some forty years before 'desertification' became an international buzzword, a discussion was held at the Royal Geographical Society entitled, 'The encroaching Sahara: the threat to the West African Colonies'. In Kenya, streams which had once been perennial had become seasonal. In Uganda, a Cambridge expedition of 1930-31 (led by Vivian Fuchs, later to become renowned as a polar explorer) reported that reduction of stream flow was responsible for lowering the levels of Lake George and Lake Edward. In 1939 the Kenya government invited a South African, Ian Pole-Evans, to tour the country, 'with the object of giving government officers some advice.' This unusually frank statement of objectives was fulfilled by making no fewer than 130 'general recommendations!'

Gullying in a valley floor (*dambo*), Malawi; the gullies lower the
water table, hence dry-season pasture is no longer available

Salinization of irrigated land, Pakistan

Early concern was not confined to Africa. Erosion committees were in
existence in the early 1930s in Sri Lanka and in the Punjab, India. In 1946
Sir Harold Glover, Chief Conservator of Forests, described erosion in the

hill areas of the Punjab, including many graphic photographs of deforestation, sheet erosion, gullying and ravines. In 1942, Frederick Hardy described
erosion in Trinidad.

Colonial governments responded to this awareness and concern by the
appointment of soil conservation officers, and in some cases, Conservation
Divisions or Services. Early conservation staff were K.J. Mackenzie in
Zimbabwe from 1936, J.H.K. Jefferson in Sudan from 1937, Colin Maher
(Soil Engineer) in Kenya the same year, and J.M. Howlett in Malawi from
1942. Conservation staff were in post in Ghana and Fiji by 1950. There
were Conservation Divisions in Kenya by 1938, and in Malawi and Sudan by
1950; in 1955 the Zambian Department had a staff of ten. In Zimbabwe, a
Department of Conservation and Extension Services (CONEX) was
founded in 1948. Its original mandate was to provide advice to (white)
large-scale commercial farms, and there was a parallel service for the
African reserves; these were later amalgamated as AGRITEX, Agricultural
and Technical Services. The notable feature of CONEX was that it integrated agricultural advisory services with conservation.

Lesotho believed in calling a spade a spade. In 1943 they established a
separate unit, Anti-Soil Erosion Measures, headed by an Anti-Soil Erosion
Officer, L.H. Collett (South Africa). In 1950 Collett's post was changed to
Soil Conservation Officer (and he received the MBE). Like many field
officers in Lesotho, his salary was augmented by a £60 horse allowance.

After independence there were initially severe setbacks. Conservation had
been associated with Colonial governments, sometimes with a degree of
coercion which made them unpopular with farmers, and not something to
which newly independent governments could give much support. This
neglect eventually changed in many countries, with recognition that land
resources need to be conserved. Conservation efforts were renewed, sometimes with the assistance of aid projects. An example is Kenya, where
conservation was neglected for the first ten years after independence. Then
in 1972, at the Stockholm Conference on the Environment, the government
of Kenya identified soil degradation as a major national problem and
requested international assistance. This resulted in a long-running Swedish
aid programme from 1974, led by Carl Wenner and with the active support
of the founding President, Jomo Kenyatta.

Conservation by earth structures

There are two ways of setting about soil conservation, one based on soil engineering or earth structures, the other known as land husbandry or conservation agriculture. Earth structures are of two basic kinds, terracing and contour bunds. In the Far East and South-East Asia, terracing might almost be called 'as old as the hills', being spontaneously constructed by farmers on sloping lands long before there was any kind of official advisory service. Outside of Asia, terracing has rarely been successful. In the 1970s an attempt was made to introduce it to Jamaica, led by a World Bank consultant, T.C. Sheng, whose homeland was Taiwan. Not many years later, all the terraces except one 'demonstration farm' had been abandoned. The more widely adopted method was a system of contour bunds and waterways. This consisted of low earth banks constructed almost along the contour, with a gentle lateral slope and ditches above them to carry the runoff to grassed waterways. This method was widely and successfully adopted in Zimbabwe. On the other hand, contour bunds were unsuccessful when introduced by the World Bank in Central Malawi, on the Lilongwe Land Development Project; farmers considered that the banks and ditches took up too much land, and they were soon planted over.

Earlier work in Malawi, beginning before the Second World War, had been more effective. Much indigenous cultivation had earlier been on mounds, leaving the soil between them open to runoff. The Department of Agriculture sought to replace these with cultivation ridges, hoed along the contour and with permanent 'marker ridges' at intervals, left under natural vegetation. This practice was adopted throughout the country, and for many years was moderately successful in checking erosion, until population increase forced cultivation onto steeply sloping land.

The Swedish aid programme to Kenya, planned and led by Carl Wenner, introduced a new kind of contour ridge called *fanya juu*, Swahili for 'throw (earth) upwards'. In the standard contour bund, soil washed down from the cultivated areas fills up the ditches. In *fanya juu* the bank is above the ditch, so soil accumulates against the bank. The drawback is that when earth inevitably collapses into the ditch, throwing it onto the bank involves heavy labour. The success of this method, judged in terms of farmers who spontaneously adopted and maintained it, was limited.

Severe, irreversible erosion in Haiti. This hillslope was covered by a
conservation programme of contour hedgerows promoted by an aid agency;
the foreground shows consequences a storm breaches the conservation works

An instructive example of failure comes from Lesotho. Some of the better
land in the lowlands was ceded to settlers, and the indigenous farmers
moved towards a cash economy, adopting European crops and technology.
Population increase led to land pressures, and by the 1920s farms were
small, leading to absence of fallows and reduced fertility. Livestock were
displaced into the highlands. When sheet erosion became widespread, the
government instituted a programme of contour bunds and ditches, which
was widely applied and became something of a showpiece for visitors.
Unfortunately, heavy storms caused breaks in the bunds. As these were

designed to direct water flow laterally, water was concentrated there. This led to the breaks extending all down the slope, forming gullies, which have since become deep and extensive.

Forerunners of a new approach

A view sometimes encountered is that the dominant method of soil conservation in Colonial times was based on earthworks, and that this was succeeded from the 1970s onwards by the new approach of land husbandry or conservation agriculture. This is by no means the whole story. Many of the features of conservation agriculture, both in approach and methods, were already known and to some degree practised in Colonial times. In 1949 Harold Tempany was commissioned to review the practice of soil conservation in the British Commonwealth. He sent out enquiries to Directors of Agriculture, from which thirteen replies were received (by no means a bad response for questionnaires). The first chapter includes an amazing paragraph which deserves quotation:

> In the words of Mr Colin Maher, Soil Conservation Officer, Kenya, 'Soil conservation can only be successful if it is related to improved husbandry, including … the maintenance of soil fertility and soil structure by all the methods known to the good husbandman. Soil-conservation methods which are not based on this (and this in Native Reserves may involve sweeping social and economic changes) are a waste of time and money and will not have lasting results.'

Colin Maher's statement is the more remarkable in that the post to which he was first appointed was Soil Engineer. He should surely be regarded as a prophet of conservation agriculture.

Tempany's review is revealing. Chapter 2 is on 'Methods of soil conservation dependent on the use of earth structures', but Chapter 3 on 'Methods … dependent on cultural practices and the use of living plant material' is of equal length. It includes descriptions of mulching, strip cropping, contoured grass strips (including Napier grass and Vetiver grass), live wash-stops for preventing gullying, protective covers, and tree planting. There is even a section on 'Contoured planted wash-stops and hedges using plants other than grasses', thirty years before their use in modern agroforestry.

'Experience in Tanganyika [and] Uganda indicates that the most successful so far tried has been *Leucaena glauca*.' Now called *Leucaena leucocephala*, this was the most widely used species in hedgerow intercropping research from the 1970s. Other present-day agroforestry species cited are *Tephrosia* and pigeon pea. 'In Ceylon, the establishment of contour hedges ... is considered an essential feature of soil-conservation practices.'

The report continues with further anticipations of the modern approach. 'Soil conservation must be integrated into systems of husbandry and not regarded as a separate end.' Soil surveys should be carried out, as a means of providing basic information:

> On the social and economic side, surveys of agricultural conditions are equally necessary [i.e. farm system studies] ... They are best organized on a team basis [interdisciplinarity] ... Means are required to test new systems in practice in the area for which they are intended [adaptive research] ... by what has been termed the 'unit-farm technique' ... based on a selected peasant cultivator and his family [on-farm research].

Tempany also refers to the possible breakdown of existing farming methods under the stress of altered circumstances, and the need for 'systems capable of indefinite continuation' — in other words, sustainability! This is a truly remarkable review, which illustrates the forward thinking to be found in Colonial territories.

Conservation agriculture

When the modern approach was first devised in the 1960s it was called land husbandry, a term used in England from mediaeval times to refer to agriculture in general. It has since become more widely know as conservation agriculture. The two terms are more or less equivalent, differing in emphasis in that land husbandry focuses on the land or soil, whilst conservation agriculture combines soil conservation with the productive function of agriculture.

Conservation agriculture refers to systems of land use and management which increase or maintain farm output, whilst at the same time conserving the soil on which agriculture depends. It thus meets the requirements of sustainability.

The approach was pioneered by two staff who spanned the period from Colonial Nyasaland to independent Malawi, Francis Shaxson and Malcolm Douglas. The titles of Shaxson's post indicate a change of emphasis: Soil Conservation Officer 1958-62, Land Husbandry Officer 1968-76. Douglas served under the latter title 1974-80, and subsequently became a leading advocate of conservation agriculture at FAO.

Robert Green, Francis Shaxson and other conservation staff had become dissatisfied with the application of the methods of CONEX to Malawi. They held a 'Land use training course' at Zomba in 1970, which became an exchange of ideas between the staff who were nominally the instructors and those attending the course. This seminal course led to the setting up of a Land Husbandry Training Centre in Zomba, and production of a *Land Husbandry Manual* in 1977. From the 1990s onwards the approach has been advanced through conferences and a string of publications by Shaxson and fellow spirits, notably Malcolm Douglas, Amir Kassam and Jules Pretty.[1]

Others who took part in the development and advocacy of land husbandry include Robert Green of Malawi, Norman Hudson of UK, Eric Roose of France, Bill Moldenhauer of the USA, R. G. Downes of Australia, and Douglas Sanders of the FAO Soils Service. Together with Shaxson, they came together to formulate the principles and practices. The conversion of Hudson to the new approach was remarkable. Educated as a civil engineer, he was employed in Zimbabwe 1951-64 as a conservation engineer, later becoming Professor of Field Engineering at Cranfield University, UK. His earlier book *Soil Conservation* (1971) was for long the standard text. The greater part of it is devoted to conservation by means of earth structures. Late in his career, in part after his retirement in 1984, Hudson became a convert to conservation agriculture, and lent his considerable authority to propagating the approach.

Conservation agriculture was founded in response to the undeniable fact that earlier methods were often unpopular with farmers. Reasons for this were firstly, the labour involved in constructing and maintaining the conservation works; secondly, the fact that no immediate benefit, in the form of improved crop yields, could be seen; and lastly, the fact that conservation

---

[1]   Malcolm Douglas died in 2009 whilst still active in soil conservation consultancy.

was often imposed from above by government officials. Given this unsatis-factory situation, conservation agriculture sought to integrate conservation into farming practices, so that increased crop yields would result right from the start, and to 'talk with farmers', to harmonize the views of farmers with those of conservationists. One way to do this was to stress the effects of conservation on improving soil water conditions, an aspect of which farmers are very aware.

Talking with farmers: Anthony Young, World Bank Sokoto
State Agricultural Development Programme, 1982

The objectives and methods of conservation agriculture are:

- *Talk with farmers*, making the first objective to increase their productivity, in the short term as well as the long; thus getting farmers to realise, 'Yes, in order to do this, we need soil conserva-tion'.

- *Maintain the physical condition of the soil.* Farmers, especially in subhumid to semi-arid zones, recognize the importance soil water--holding capacity.

- *Maintain soil organic matter status and biological activity.*

- *Maintain a continuous soil cover*, either by a dense crop layer or by keeping crop residues or leaf litter on the surface.

- In some cases *minimum tillage*, avoiding compaction of the surface layer through excessive ploughing or hoeing.

- *Maintain soil nutrient status*, through a combination of organic matter and biological activity, combined with judicious use of limited quantities of fertilizers.

- *Avoid erosion*, preventing the physical loss of soil material by sheet erosion and gullying.

Thus in the older approach to soil conservation, avoidance of the physical loss of soil matter by sheet erosion or gullying was the primary objective. In conservation agriculture, checking erosion remains essential but is not made an explicit objective; one published account called it 'Soil conservation by stealth'.

The rationale, approach and methods are set out in *Land Husbandry: a framework for soil and water conservation* (1989), a clear and concise account (although strangely, most of the photographic illustrations show conservation through earthworks).

Minimum tillage

The practice of minimum, or zero tillage, arose separately. In the 1970s, farmers in southern Brazil suffered disastrous soil erosion because they bared their soil to the elements and destroyed its structure by ploughing. Research came up with three recommendations: first, break up compacted layers to increase infiltration and storage of rainwater; secondly, retain crop residues (as opposed to burning them) to provide a buffer against rain splash; and thirdly, seed directly through the mulch to ensure the least disturbance of the surface cover and the soil beneath. There are big savings in cost of machinery, although chemical treatment may be necessary to get rid of weeds which had formerly been destroyed by ploughing.

These ideas caught on, and were included under the banner of conservation agriculture. Minimum or zero tillage was found to achieve 30-90% reduction in runoff, with a corresponding reduction in soil erosion. In 1974 conservation agriculture with minimum tillage covered only some 2.8 million hectares world wide; by 2004 this had reached 72 million hectares, predominantly in developed countries, with coverage of some 60% of arable land in South America, 40% in Australia and 24% in North America.

Minimum or zero tillage is largely carried out where mechanized farming is the norm. The practice has spread to some small farms in China and India. Ironically there has been little takeup in Africa, where conservation agriculture began.[1]

Agroforestry and conservation agriculture

Farmers have been planting and managing trees on farms for thousands of years, although the name 'agroforestry' was not coined until 1977 with the establishment of the International Council for Research in Agroforestry (ICRAF) based on Nairobi.[2] One of their earliest activities was to review the potential of agroforestry for soil conservation. The sponsors of this probably thought of it as control of erosion but it was taken in its wider sense to include maintenance of soil fertility. At the time it was difficult to assess this potential owing to the lack of experimental results from agroforestry systems, and the resulting book, *Agroforestry for Soil Conservation* (Young, 1987) was perforce based mainly on data from related land use systems. Later this situation had been transformed, and the second edition, *Agroforestry for Soil Management* (Young, 1997), retitled and largely rewritten, was able to draw on ten years of research specific to agroforestry systems.

Trees on farms have an evident potential for protection and improvement of soils. The canopy protects from raindrop impact, leaf litter provides a ground cover, its decomposition supplies soil organic matter, and if nitrogen-fixing trees are used there is an input of nitrogen. Root studies and modelling of the organic matter cycle showed that the growth and ongoing decay of root systems makes a major contribution to soil organic matter.

---

[1] I have to confess to a blank spot when it comes to minimum tillage. The statistical evidence in its favour is strong, and its advocates strongly support wider adoption. However, having never seen it in practice, I am not in a position to form a view.

[2] Now the World Agroforestry Centre.

In 1949, long before the start of modern scientific agroforestry, D.G.B. Leakey in Kenya had advocated the planting of fodder trees along contour strips to check runoff and erosion. In the Philippines and at the International Institute of Tropical Agriculture (IITA), Ibadan, Nigeria, research was conducted into a system of planting low hedges parallel to the contours, correctly called hedgerow intercropping but more widely known as alley cropping.

The system was taken up, along with more wide-ranging studies of agroforestry for soil conservation, by Paul Kiepe and myself, working at ICRAF. Hedgerow intercropping was found to be effective in checking erosion. This was not, as had at first been assumed, by the hedges acting as a barrier to runoff. Infiltrometer measurements showed that the main effect was that the roots of the hedgerows increased the rate of infiltration, thus reducing runoff. A side-effect is that the necessary prunings from the hedgerows can provide soil cover and inputs of organic matter and nutrients. From a technical viewpoint, and on experiment stations, hedgerow intercropping was a highly successful system.

The system was widely propagated but unfortunately, after the initial enthusiasm, was no more successful than earth structures in achieving spontaneous adoption. Farmers basically do not want to cover their fields with hedges. Where they have been persuaded (sometimes paid) to do so, after a few years the breaks in hedgerows caused by storms were not maintained.

Other agroforestry systems, based on canopy protection and the maintenance of a continuous ground surface cover of tree leaf mulch, combined with outputs from the trees themselves (fuelwood, fruit, fodder) have been more successful, and have contributed to the range of tools available to conservation agriculture.

Land degradation

When it came to justifying expenditure on soil conservation, problems arose in demonstrating its severity of erosion. Speculative and highly exaggerated statements were made, such as claims of erosion rates of 30-40 tonnes per hectare per year, or that one third of all cultivated land had been irreversibly destroyed. 'Desertification' became a buzzword in project proposals. What was needed was to identify degradation in relative terms: which regions are

most affected, and is it related to population pressure or poverty? In particular, there was a need to identify what were called 'hot spots', for which
action was most needed.

Direct measurement of erosion in the field is difficult other than on small
fenced plots, so resort was made to the use of modelling. This was founded
on the Universal Soil Loss Equation (USLE), which gives the predicted rate
of erosion, in tonnes per hectare, as the product of five factors: rainfall
energy, soil resistance, slope angle and length, crop cover, and conservation
practices. Calibration was based on many thousands of plot-years of experimental data. Having obtained predicted erosion without conservation measures, rates were reduced to levels called 'tolerable erosion' (assessed by some
highly questionable assumptions on rates of natural soil formation). When
applied to African conditions the USLE gave unrealistic results, effectively
barring cultivation from all but gentle slopes, and a modified version, the
Soil Loss Estimation Model for Southern Africa (SLEMSA) was developed
in Zimbabwe.

Having completed the first *Soil map of the world* in 1978, FAO proposed to
follow it with a world map of soil erosion, to be based on the USLE. To
discuss this project a meeting was called, Methodology for Assessing Soil
Degradation. This reached the conclusion that to apply, on a world scale, an
equation based on small experimental plots was unrealistic, and unusually
for such consultations, recommended that the project should not go ahead.

The first attempt to gain a world view was the Global Assessment of Soil
Degradation (GLASOD), carried out by ISRIC[1] in conjunction with UNEP
and FAO. As so little reliable quantitative data were available, comparative
assessment of the severity of degradation was made by the following
method:

- Establish soil mapping units, based on the FAO *Soil Map of the
  World*.

- Define types of degradation (water erosion, wind erosion, soil
  chemical and physical degradation, and salinization and waterlog-

---

[1]   Formerly the International Soil Reference and Information Centre, now World
      Soil Information, Wagengingen.

ging), and degrees of severity (light, moderate, strong, extreme) based on effects on agricultural productivity.

• Ask national collaborators in each country to assess the percentage of each mapping unit affected by each degree of severity.

The results give estimates of the proportion of usable land (excluding non-agricultural land, e.g. deserts, mountains, ice-covered) which are affected by different degrees of degradation. By tacit assumption, this is the summation of all past degradation. Thus for Africa, for the 1663 million hectares of agricultural land, 30% is subject to some degree of degradation, and 8% to strong or extreme degradation. These figures are somewhat lower for Asia, and substantially lower for South and Central America. The world study was followed by a more detailed assessment for Asia (inevitably ASSOD).

The outputs are essentially the summation of estimates by local staff, some 200 members of national soil surveys or conservation departments. It can be described as a compilation of expert judgements. Because they are given in numerical terms, as percentages, the results give the appearance of being quantitative.

These results are certainly better than previous sweeping generalizations, and were widely quoted over the succeeding 20 years. However, they are rightly criticised as being founded on what are at base, subjective estimation. It is not beyond possibility that in some cases degradation severity was over-estimated, with a view to attracting international support.

The next attempt also originated at ISRIC. The concept came from Michael Schaepman, Professor of Remote Sensing, and David Dent, Director. The objective was to provide a framework for the chapter on Land in UNEP's 2004 Global Environmental Assessment, for which Dent was the coordinating author. Since 1981 the Advanced Very High Resolution Radiometer carried by NOAA weather satellites had been scanning the globe daily at 2 km definition. By good fortune, the ratio of red to near-infrared radiation was found to provide a good measure of the photosynthetic capacity of the plant cover. This ratio was called the Normalized Difference Vegetation Index (NDVI) and used as a proxy assessment of land degradation. Drought effects were masked using a computation of rain-use efficiency. This massive study, intended to be an improvement on GLASOD, was originally called the Quantitative Global Assessment of Land Degradation and Improvement.

This caught the attention of an FAO project, Land Degradation Assessment in Drylands (LADA). Workshop preparation and planning had begun in late 2000, with the first draft project proposal the following year. The project was implemented in 2006, with national participation in six countries: Argentina, China, Cuba, Senegal, South Africa and Tunisia, with spin-offs to a larger number of countries not formally part of the project. A new methodology for degradation assessment was developed, manuals for local degradation assessment prepared, regional training workshops held, and the results checked in the field. After five years of activity the project was terminated in 2010. As international projects go, LADA can be considered successful.

LADA needed a means of assessing land degradation on a global scale, and the ISRIC study appeared to supply this. FAO therefore took it up, renaming it Global Assessment of Land Degradation (GLADA). This had the effect removing a limitation to LADA, which was confined to drylands (very broadly defined). GLADA defined land degradation as decline in ecosystem function and productivity, and thus employed used plant growth, measured by satellite remote sensing, as an index of the productive capacity of land. This is taken as a proxy assessment of land degradation or, where a later value is higher, land improvement.

Differences between the two projects directed at measuring land degradation on a world scale are:

- GLASOD estimated the sum of historical degradation up to the present, GLADA the recent rate of degradation.

- GLASOD gave results for type of degradation (water erosion, wind erosion, vegetation degradation, salinization, etc.), GLADA only the net effect on land productive capacity.

- GLASOD was based on direct but subjective estimates, GLADA on an objective, quantitative but indirect method (its authors call it a proxy assessment).

As would be expected, the two assessments differ considerably. GLASOD lent support to popular supposition, that the most degraded areas are in drylands, the semi-arid zones, notably the sahel belt of west Africa.

GLADA cast doubt on this view. 'The results are very different from previous assessments which compounded what is happening now with historical land degradation' states the summary report. The areas most severely affected by ongoing degradation are found to be Africa *south* of the equator; a region of south-east Asia including Indo-China, Myanmar, Malaysia and Indonesia; south China; and, surprisingly, parts of eastern Australia. Over 50% of the land area is said to be degrading in 19 countries: nine in Africa, seven in Asia, two in central America, and New Zealand. The highest values of degrading land are for Swaziland (95%) and Thailand (60%). Unexpected features are that many of the most affected regions have humid, rainforest, climates, e.g. Zaire, Congo, Indonesia and Malaysia. Contrary to widespread supposition many semi-arid regions, notably the sahel belt of Africa, show a positive trend in net primary production, inter-preted as mainly due to changes in climate.

Which is the most severely eroded country in the world? The maps based on the two international assessments of land degradation show that the most severe erosion is in no way country-specific but occurs as localized areas, often strips of sloping land that have been occupied for cultivation. There is one country, however, in which these reviews coincide with my personal experience. The GLASOD maps show almost the entire area of Haiti as class Wt3.5/Wd3.3, severe water erosion with mass movement. Haiti contains much sloping land from which the original forest cover has been almost entirely removed for cultivation. The result has been extensive and severe erosion, leaving large areas almost entirely without soil of any depth, leading to abandonment by farmers. Once erosion has reached this degree of severity the results are irreversible.

An unexpected finding arising from the method employed in GLADA, measurement of net primary production from satellite imagery, is that since 1982 there has been a substantial and steady global increase in rates of plant growth. This feature is found both north and south of the equator, and in all continents. A plausible explanation is that plant growth is responding to rising levels of atmospheric carbon dioxide.

Only by going into the field, examining and monitoring soils and vegetation, and talking to farmers, can the reality of these attempts to map land degradation be evaluated. There is no substitute for the study of soils in the field.

14

RETROSPECT: THE SURVEYORS

BUILDING UPON PIONEERING work before the Second World War, the era of reconnaissance surveys which followed it led to a massive advance in knowledge. Studies carried out within the Colonial service[1] were continued after independence by the Land Resources Division. Both were supplemented with project surveys by consultant companies. No one today need start a rural development without a framework of information on land resources: the kinds of land that exist, and where they are to be found. It is instructive to look back on what was achieved, how much it fell short of the intentions, what mistakes were made, and the legacy for the present day.

First, let us look at those who carried out the work. There were surprisingly few of them, but the Colonial Service as a whole did not take on staff without good reason, entrusting its sparse personnel with considerable responsibilities. Before 1940, de facto soil surveyors numbered perhaps 15: the six pioneers of Chapter 2 together with a scattering of soil chemists, ecologists and agricultural officers who saw a need to undertake local surveys.

Expansion in staffing began soon after the end of the Second World War. Indeed, C.F. Charter was in post in Ghana from 1944, and S.P. Raychaudhuri in India the same year. Brian Anderson was recruited to Tanzania from 1947, and John Coulter to Malaya the following year. The main post-war expansion in staff began with the 'Cocoa Survey' in Western Nigeria from 1951, and C.F. Charter's recruitment drive in Ghana, where from 1953-57 he took on seven officers. Peak numbers of expatriate surveyors in the Colonial Service were reached soon after that. Summing the soil surveyors and ecologists given in Colonial Office staff lists for

---

[1]   In full, His/Her Majesty's Overseas Civil Service, or H.M.O.C.S.

1950-1970 gives 56, to which should be added 24 who worked for the Regional Research Centre in the West Indies. Following the main period of Colonial independence, 1956-70, expatriate numbers dropped and there was a delay before local staff could complete their training. Some of the returning British staff were taken on by the Land Resources Division, which expanded in numbers through the 1970s, and by the early 1980s had reached 18 staff in the areas of soils, ecology and land use planning.

Adding together Colonial staff, the Land Resources Division, eleven with Huntings, at the peak ten with ULG, some with other consultant firms, and a few from international agencies gives a total of 120 soil surveyors and ecologists working in British territories in the period 1950-75.[1] A count of the soil surveyors listed in the index to this book gives about 150. This is certainly an underestimate. The total throughput must have been 150-200 or more, of whom less than half would have been in post at any one time. To these can be added the expanding staff of the Indian Soil Survey. However, detailed country studies conducted for the World Soil Survey Archive and Catalogue (WOSSAC), together with personal communications from their authors, reveal that substantially more individuals joined surveys for short periods.

Considering the magnitude of the tasks accomplished in some fifty countries, these are remarkably small numbers. As a student essay rightly if ambiguously expressed it, 'In the tropics, Soil Surveyors are often spread thinly over the ground'.

From other nations

Surveyors originating from outside the United Kingdom played an important role with Colonial governments, consultant companies, and international surveys of British territories. Some adopted British nationality. The most numerous were from the Netherlands, many having studied at the Agricultural University of Wageningen. They included Jacobus Andriesse, Robert Brinkman, A.H. Buddingh, Herman Haantjens and Pieter Bleeker (the last two both with CSIRO Australia in Papua New Guinea), P.D. Jungerius, Willem Sombroek (heading Dutch aid to the Kenya Soil Survey),

---

[1] Having reached this number, I turned to a letter from the late George Murdoch. Out of his head he had listed 120!

Jan de Vos, and Iraneus Ysselmuiden. The authors of that fine synthesis of West African environments, *Agricultural Ecology of Savanna* (1978) both hailed from abroad, Jan Kowal from Poland and Amir Kassam from Zanzibar. Stanislav Radwanski from Poland had the distinction of taking part in three major reconnaissance surveys, with Colonial governments on Ghana (1951-1956) and Uganda (1956-1959), and with FAO in Pakistan (1960-1963.)

Eberhard Brunig became Forest Conservator in Sarawak in 1954. A colleague told him that wartime service counted for salary increments, and he approached the administrative officer to find if his army service would qualify. 'Certainly', he was told. 'Does it matter if it was in the German army?' After gaining nine years' experience in the Colonial Service, Brunig went into an academic career, becoming Professor of Tropical Forestry in Hamburg.

Women

It is usual in writing of a profession to give special attention to the role of women, but in the present case this is difficult. The Colonial Service as a whole employed very few women at professional grade, other than in secret-arial work, nursing, education and other roles in which, at the time, women's skills were thought to be suited. True to long-standing practice, the staff lists usually omit the date of birth of female officers.

Male dominance was most strongly marked in forestry, and in topographic, geological, and soil survey. A small number of female ecologists can be found in Colonial staff lists. Uganda appears to have made an early attempt at positive discrimination, in 1956 creating posts of 'Librarian (woman)' and 'Taxonomist (woman)'. The latter was filled by Miss M.E. Griffiths in 1957 but, sad to relate, was recorded as vacant the following year. Similarly Miss M.E. Parry served as one of eight botanists from 1965 to 1967; a colleague recalls that she married another botanist. By 1972, localization had brought about the further innovation of appointment of a married woman, one of four botanists being Mrs E. Rubaihayo.

In the area of soils, however, women staff are few. A photograph of the First International Agro-Geological Congress, (the precursor of interna-tional soil science congresses), held at Budapest in 1909, shows seventy participants dressed in frock coats and bowler or top hats, and one single woman, her skirt brushing her shoes. George Murdoch, in sending a list of 120 surveyors includes three women, Julie Jones, Sally Sutton and Penelope Smith (Rhodesia), but I have failed to trace them. The shining exception in Colonial times is Helen Brash, whose efforts to overcome male prejudice against travelling on field survey have been related in Chapter 5. The Land Resources Division operated in later years, when male dominance of the professions was becoming less pronounced. On the Central Nigeria project of the early 1970s, out of a team of twenty staff, two are women: Judith Jones as Geomorphologist, and Mary Alford as Ecologist.

Susan O'Farrell began her love of the tropics, like many young graduates, with the British charitable organization, Voluntary Service Overseas (VSO) in Nigeria. She then joined Huntings and worked as Soil Surveyor on the South Chad Basin study, Nigeria. Later in Papua New Guinea she widened her interests to gender issues and poverty, for which she was awarded the MBE.

Education

There were three basic educational routes for British expatriate soil surveyors. The first way was to start with a degree in geography or geology, then either take a postgraduate course in soil science or learn the necessary chemistry as best they could. Of those for whom I have information, this is the largest group. Cambridge and Oxford geographers take the lead, but this is affected by the selective mode of recruitment of C.F. Charter (natural sciences, Cambridge). Hugh Brammer (geography, Cambridge), Maurice Purnell (geography, Oxford), Clifford Ollier (geology with geography, Bristol), Ronald Paton (geology, Durham) and Anthony Smyth (Geology, Cambridge) are representative of this group.

The second route is to read chemistry or agricultural chemistry, take up a post of soil chemist, then learn geology, geomorphology, ecology and climatology from colleagues and in the field. This is a tall order, yet some of the most distinguished soil surveyors managed it. The most notable was Geoffrey Milne (chemistry, Leeds), but we must suppose his field awareness gained much from his wife Kathleen (geography, Aberystwyth). Others who followed this path include Graham Higgins (agricultural chemistry, Aberdeen) and Richard Webster (chemistry, Sheffield); Alun Jones embarked on a mathematics degree but when this was interrupted by dealing with explosives on war service, he felt it would aid his self-preservation if he changed to chemistry.

The third and most obvious route was via a first degree in agriculture, with postgraduate specialisation in soil science. There were more Departments of Agriculture in British universities at the time than there are now, and soil surveyors notably came from Oxford, Reading, Edinburgh, Aberystwyth, Wye College London, and Cambridge.

The important postgraduate topping up in soil science and survey, often supported by government scholarships, could be found at Oxford, Reading, Newcastle, Aberystwyth, and the Macaulay Institute at Aberdeen. It might also take the form of a secondment to the Soil Survey of England and Wales as a kind of field apprenticeship, either to its headquarters at Rothamsted or with one of its surveyors in the field. Knowing that I was to be posted to the African Rift Valley, the Ministry sent me to the Vale of York (!) to learn from Alan Crompton. Overseas, the most frequent assignment was to the Diploma in Tropical Agriculture (DTA) at the Imperial College of Tropical Agriculture, Trinidad, where postgraduates could benefit from the wisdom of Frederick Hardy. Very few British graduates benefited from the incomparable training in remote sensing provided by the International Training Centre for Aerial Survey and Earth Sciences (ITC) at Wageningen, the Netherlands. Among the most fully trained British surveyors were John Hansell, with the agriculture degree and soil science diploma from Oxford, followed by the Trinidad DTA, and the ecologist Ian Langdale-Brown, with botany at St Andrews, postgraduate ecology at Oxford, then training with Trapnell in East Africa.

The local staff who took over from expatriates generally began with an agriculture degree in their own country or region, followed by a scholarship overseas. Over many years the Dutch government generously supported Third World students at ITC. The British brought in students to their Masters degree courses, or sometimes for the three years or more of a PhD. A few went to the Soviet Union, a long-drawn out training as the first year was spent learning the language. The most prolifically qualified was the Zimbabwean, Kingston Nyamaphene who, before his government would appoint him as soil surveyor, had to prove his ability by gaining no less than four qualifications: a BA in geography from London, Certificate in

Education, Rhodesia, Diploma in Remote Sensing from ITC Netherlands, and MSc in agronomy from Cornell, USA. He later added a PhD in soil science from Aberdeen.

To complete a PhD before taking up a soil survey post was not common. Harry Vine took one in chemistry before going out to ICTA. I completed one in the geomorphology of slopes, a somewhat non-applied branch of the subject since changes take place over a time-scale of millions of years. More commonly, staff already in the field felt that this qualification would improve their career prospects, and worked on an external student basis for many years. Some examiners would comment that there was 'nothing original' in a thesis based on field survey, calling for the addition of some superfluous statistical exercise, but failing to appreciate that to work out for the first time environmental interactions in the field is always original work. A soil surveyor for Huntings, Karl Kucera, originated from Czechoslovakia. He began his student life by dropping Molotov cocktails (petrol bombs) into Soviet tanks invading Prague. Working for Huntings and registered as an external student he took great pride, for the honour it brought to his father, in a University of Anglia (Cambridge) PhD. Christopher Panabokke's path was from a chemistry degree in his own country, Sri Lanka, to a soil science PhD at Australia's leading agricultural university, Adelaide.

For originality, we turn again to Charles Wright. With some difficulty, he was persuaded to submit a PhD to the University of Leeds. Having had it accepted, after a celebratory sherry he withdrew it.

Career paths

In the years before the Second World War, Colonial Service was normally a job for life. There were promotions from 'Officer' to Senior and Principal levels, or for a few, to Deputy Director or Director of Agriculture, and quite often transfers from one territory to another. This 'no change' career path applies to all of our six pioneers. Hardy stayed as Professor at ICTA and retired there, Milne (East Africa) remained in soil survey and died in office, Hornby reached the level of Acting Director of Agriculture, Malawi, and retired in his country of adoption, Martin (Sierra Leone) became Director of Agriculture quite early in his career, Joachim towards the end of it. Trapnell, after spending what for many would have been the greater part of a career in Zambia, moved to a British government post of training in ecological survey in Kenya, and well after retirement age founded his own institute at Oxford University.

Others who started shortly before and after the war stayed in soil survey. These included C.F. Charter, who died in office as Head of the Ghana Soil Survey, Alun Jones (West Indies), Charles Wright, whose tropical career was mostly based on the New Zealand Soil Survey, and Bill Panton in Malaya.

Those joining the Colonial Service in the peak period of recruitment, the 1950s, found their expected careers shortened or terminated by political change. Quite a number stayed on for a few years after their country's inde-pendence, before handing over to local staff. For this generation, there were five main career paths: the Land Resources Division, consultant companies, international institutions, British institutions, and the academic field. There must surely have been a few who decided that a career in land resources was not for them, but only one case has been brought to my attention. Anthony Mitchell, 'decided he would like to try something different' and spent four years working for a financial services company — but that was after twenty-seven years in field survey.

The Land Resources Division (LRD) was an obvious career path. It was in some sense a successor to resource surveys by individual countries, and naturally sought to recruit staff who already had overseas survey experience. Some who moved from Colonial Service (including the Colonial Pool) to LRD were Douglas Carroll, David Lang, Bryan Acres (who died in post), Alan Stobbs, John Hansell and David Wall; Carroll later joined the Soil Survey of England and Wales. Michael Bawden, after a long period with LRD, moved to a post with the British Government's Overseas Develop-ment Agency. Anthony Smyth's distinguished and varied career has been described in Chapter 6.

Robert Ridgway began in 1968 as a topographic surveyor with the Direct-orate of Overseas Surveys, retrained in remote sensing at Reading Univer-sity, joined FAO as a land use planner, then spent the later part of his career with LRD, by then the Natural Resources Institute. In completing a project as Land Reform Adviser in Namibia (2005-10) he became the last working professional in area of land resources with the former Land Resources Divi-sion, which by then had become the Natural Resources Institute.

Others stayed with LRD for the greater part of their careers, John Bennett and Anthony Mitchell after apprenticeships with Huntings. The extreme case is Martin Brunt, who was the founding staff member of the Division in 1956 and remained with them for the rest of his working life. The CVs of long-serving LRD staff such as Brunt, Bennett and Ridgway make impressive reading, with work experience in upwards of twenty-five countries. Some might consider that such frequent moves, the normal pattern since 1960, meant that they never acquired the deep knowledge of a country of long-serving staff in earlier periods. Against this, it meant that jobs got done where they were called for, and experience gained in one country was transferred to another.

Consultant companies were also pleased to take on staff with previous experience, both for what they knew and to show this on applications for contracts. Stanislav Radwanski and Brian Anderson followed this route, whilst George Murdoch moved to Booker Agriculture. Some remained with consultants for all or most of their careers. Examples are Vernon Robertson, founder and subsequently Managing Director of Hunting Technical Services, Stan Western, also with Huntings, and Rick Landon with Booker Agriculture. David Parry moved from Huntings to Mott Macdonald, reaching project manager level, and Keith Virgo from Huntings to Land and Water Management Ltd.

Two sets of international organizations took on Colonial and LRD staff: United Nations Agencies (FAO and the World Bank), and the international agricultural research centres (mainly but not entirely belonging to the cumbersomely named Consultative Group on International Agricultural Research (CGIAR). Graham Higgins spent sixteen years in Nigeria starting as soil surveyor, moved into land resources research, and ended as Acting Vice Chancellor of Ahmadu Bello University; then joining FAO as a field officer in Pakistan, he became successively Head of the Soils Service and Director of the Land and Water Development Division. Others to join FAO were Maurice Purnell, for many years in charge of land evaluation activities, Anthony Smyth, John Harrop, Hugh Brammer (always as a field officer, never in headquarters), and Amir Kassam, who for many years was their near-permanent consultant on agroclimatology.

To represent land resources in the World Bank, other than in the context of forestry, has always been an uphill task. John Coulter, originally with the

Colonial Service in Malaya, spent his last six years of full-time employment as Agricultural Research Adviser to the World Bank, and Ian Hill moved to the Bank from LRD. Of the international research centres, Peter Ahn spent a period with the ill-fated International Bureau for Soil Research and Management (IBSRAM), whilst I joined the International Centre for Research in Agroforestry (ICRAF) early in its existence, leaving when it was ten times its former size and had gained acceptance by the CGIAR. David Dent, after a period in academic work and subsequently in Australia, became Director of the Dutch-based organization, World Soil Information (formerly ISRIC).

The national institutions which took on land resources staff were, with a few exceptions, those with an international remit. Herbert Greene, for many years soil chemist in Sudan, performed stalwart work as the UK's Tropical Soils Adviser, based on Rothamsted. Richard Webster, ex-Zambia, also joined Rothamsted Experimental Station, was for many years editor of the *European Journal of Soil Science*, and most unusually for an Englishman, spent time in a francophone institute, as Director of Research at the Institute Nationale de la Recherche Agronomique (INRA). Barry Dalal-Clayton, originally in soil survey in Zambia, joined a non-governmental organization, the International Institute for Environment and Development (IIED). David Eldridge followed a varied career path, including Land Use Planning Officer, Zambia, Soil Surveyor, Botswana, the Soil Survey of England and Wales, the University of East Anglia, finally becoming an editor for CAB International.

Many Colonial staff followed the academic path. Indeed, it may be that some had taken up a post involving field survey with the intention of using it as a stepping stone towards university work. Former soil surveyors moved into two areas, soil science and geography. Some moved to Third World universities, such as Peter Ahn to Nairobi where he became Professor of Soil Science, Harry Vine to Ibadan, and Gordon Anderson to Makerere. Others moved to Britain, like Roy Montgomery at Newcastle. Among those who moved into geography departments were Sandy Crosbie, Professor at Edinburgh, and Colin Mitchell, becoming a specialist in remote sensing at Reading. I was a founding staff member of what was to become the leading School of Environmental Sciences in the UK, at the University of East Anglia. The former ecologist in Uganda, Ian Langdale-Brown, acquired distinction at the Department of Natural Resources, University of Edinburgh. Ronald Paton joined the geography departments successively at the new University of Sussex and Macquarie University, Sydney, and Clifford Ollier followed a varied route as a geomorphologist both in Britain and Australia.

Soil surveyors of local origin naturally did not need such alternative career paths. Most remained with their national soils research organizations, with changes of emphasis from survey to wider aspects of soils research. An exception was the Zimbabwean Kingston Nyamaphene, who after a spell as Soil Surveyor 1978-81 led a most distinguished career in academic administration: Dean of Agriculture at Fort Hare and Deputy Vice Chancellor at Vista University, both in South Africa, and Dean of International Programs at the State University of New York. Henry Obeng of Ghana moved into a consultancy organization. A few were sought out by international institutions, for example the Sri Lankan Christopher Panabokke who moved into the International Water Management Institute. Many of the earliest national staff to be appointed became Directors of their national soils research institutes.

Honours

It was not uncommon for the British to reward service overseas through the honours system.[1] MBEs were awarded to Arthur Hornby, George Murdoch, Susan O'Farrell, and Peter Brown, the Malawi agronomist who contributed to soil surveys. The list of OBEs is longer, including William Allan, Hugh Brammer, Graham Higgins, Anian Joachim (Sri Lanka), Amir Kassam, Jan Kowal, William Panton, Vernon Robertson, Francis Shaxson, Colin Trapnell, and Ian Twyford. Frederick Hardy was awarded a CBE, and Charles Pereira became Sir Charles, Knight. Michael Bawden stood on the rung of a different ladder, with a CMG.

For his work in Sri Lanka over more than thirty years, Anian Joachim was awarded a Coronation Medal in 1953, and his successor, Chris Panabokke, received his own country's highest scientific honour, the Presidential Vidya

---

[1]   MBE, OBE, CBE, KBE: respectively: Member, Officer, Commander, and Knight Commander of the Order of the British Empire. KBE is usually called Knight. CMG: Commander of the Order of St Michael and St George.

Jothi award. Francis Shaxson received the Hugh Hammond Bennett award of the US Soil and Water Conservation Society.

The highest British scientific honour, Fellow of the Royal Society (FRS), was received by Charles Pereira and three others who, from academic bases, contributed much to knowledge of land resources: the climatologist Howard Penman, and soil scientists Dennis Greenland and Peter Nye.[1] Richard Webster had the unusual distinction of having an award of the International Union of Soil Sciences named after him, the Richard Webster medal for advances in the application of statistics to soil science.

For the most highly honoured in formal terms, Sir Charles Pereira has one peer, who like him combined dedication and distinction with extremely long service. This is Hugh Brammer, for whom recognition came from four sources: the OBE from the British government, the Presidential Gold Medal from Bangladesh, FAO's Sen Award for outstanding service as a field officer, and in his eightieth year, the Royal Geographical Society's Busk Medal.

Rewards

The financial rewards of a career in the Colonial Service were never great. In earlier days there was job security and a likely advance to Senior or Principal level. Following independence, some expatriates received lump-sum compensation for loss of office, known as the 'golden bowler' (hat). For most, the British government (not without extended and persistent lobbying) took over cost-of-living indexation of pensions.

It is clear, however, from all who have written about their experience, that the nature of the work itself was the main incentive and reward. There were two aspects, scientific and humanitarian.

The first reward was the experience of field survey, the excitement of being able to study landscapes about which no one had written before. For the true geographer — and this includes the soil chemists who saw the need for

---

[1]   There is no Nobel Prize for Agriculture. The equivalent is the World Food Prize, founded by the pioneer of international agricultural research, Norman Borlaug. This was awarded to a tropical soil scientist, Pedro Sanchez (USA) in 2002.

field survey — there is no experience quite like scientific exploration: going to a new area, recording the nature of its environment and landscapes, and working out how these had come into being. Cliff Ollier and Hugh Brammer speak for most, if not all, of their colleagues in saying how fortunate they regard themselves in having been able to spend their working lives in what they liked doing best.

The second aspect, common to all Colonial servants, was the opportunity to be of service to the peoples of Third World countries. This is clear from the writings and statements of many. To give a few examples, Colin Trapnell, assigned to conduct an ecological survey, directed great effort to working out how to understand the agricultural practices of the peoples of Zambia. Hugh Brammer was totally dedicated to the welfare of Bangladesh. Charles Wright, completely oblivious to physical discomfort, would spend long periods among the agricultural and forest-dwelling peoples of the countries he surveyed. John Harrop writes of his work in Tanzania, 'I derived the greatest satisfaction from identifying, studying and trying to find solutions to practical problems, even if success was sometimes elusive.' A Christian (Quaker) conviction underlay the dedicated work of Malcolm Douglas in helping farmers through better land husbandry.

Those field surveyors who turned to academic careers were able to pass on their field experience to others. The latter included British undergraduates and postgraduates, and, increasingly with time, students from overseas, usually at postgraduate level and sometimes already in the service of their governments.

Taken as a whole, during their time in the field and in their subsequent careers, these were a remarkable band of people. Why should the profession of soil surveyor have led to such distinguished careers? From the start, they showed a strong drive towards finding out about landscapes in the field. This gave them a foundation of knowledge about the environment, the nature and distribution of soils, plants, landforms and water. Based upon this experience they did not need to 'find' topics for research, these had been set by their field experience. In development work with international agencies, they knew about the reality of problems faced by the field staff.

Reunion of soil surveyors, Royal Geographical Society, 2010; from left Maurice
Purnell, Helen Sandison (née Brash), Hugh Brammer, Anthony Young

15

RETROSPECT: THE SURVEYS

SURVEYS DURING THE PERIODS reviewed fall into three groups, reconnaissance, project, and detailed. The regional accounts have focused on national reconnaissance studies, but in all countries a large number of more intensive studies, for specific development purposes, were being carried out at the same time.[1]

At reconnaissance level, by 1950 Zambia had been covered by Trapnell and Clothier's ecological survey, which demonstrated the close correspondence between soils and vegetation. The three East African territories, Kenya, Uganda and Tanzania, had been surveyed on a uniform basis by Milne, who added the invaluable approach of using soil catenas as mapping units. In both these cases, the problem of plotting boundaries without air-photograph cover was overcome by field traverses of vast length. Pioneering work in other countries was limited in two ways. There was either a sketchy basis of mapped boundaries, as in Martin's work in Sierra Leone, and Hardy's 'Grey Book' surveys in the West Indies; or a limited coverage, as in Joachim's regional surveys in Sri Lanka and Hornby's map of Central Malawi.

A transformation of the state of knowledge began soon after the end of the Second World War. Its end came about progressively, but a working date of 1970 serves to separate two groups of surveys: those carried out during the pre-independence period, and those initiated after countries became independent. The table below summarizes progress in national reconnaissance surveys in thirty-five countries.

---

[1] Sources for surveys referred to in this chapter are found in the Bibliographic entries of the respective countries.

National reconnaissance land resource surveys. For explanation of terms, see text. Some smaller island territories are excluded. RRC=Regional Research Institute. LRD=Land Resources Division/Development Centre.

Country	Institutions	Coverage	Years (publication)
Before 1950			
Kenya, Uganda, Tanzania	Government	Full	1936
Malawi	Government	Part	1938
Sierra Leone	Government	Full	1926
Sri Lanka	Government	Part	1945
Sudan	Government	Limited	1930s
West Indies	University	Full	1936/1947
Zambia	Government	Full	1937-1943
1950-1970			
Belize	Government	Full	1959
Brunei	Consultants	Full	1969
Fiji	Government	Full	1965
Ghana	Government	Part	1961-1967
Guyana	FAO	Full	1965
Hong Kong	University	Full	1960
India	Government	Part	1947 ongoing
Jamaica	RRI	Full	1961-1971
Lesotho	LRD	Full	1967-1968
Malawi	Government	Full	1965-1971
Malaysia: Malaya	Government	Full	1958-1970
Malaysia: Sarawak	Government	Full	1962-1966
Nigeria (W. Region)	Government	Part	1962
Papua New Guinea	CSIRO Australia	Full	1964-1970
Samoa	Government (NZ)	Full	1963
Seychelles	LRD	Full	1968
Sri Lanka	Government/Canada	Full	1962
Sudan	Consultants	Part	1950-1970
Swaziland	Government	Full	1963-1970
Tanzania	Government	Part	1967

Uganda	Government	Full	1959-1962
West Indies (other than Jamaica)	RRI	Full	1958-1967
Zambia	Government	Full	1970
Tanzania: Zanzibar	Government	Full	1955
Zimbabwe	Government	Full	1955-1960

Partly post-1970

Bangladesh	FAO/Government	Full	1965-1977
Botswana	LRD	Part	1966-1972
Gambia	LRD	Full	1969-1977
India	Government	Part	Ongoing
Kenya	Government/Netherlands	Full	1980
Malaysia: Sabah	LRD	Full	1975-1976
Myanmar	FAO	Limited	1972
Nigeria (regions other than West)	LRD	Part	1966-1979
Pakistan	Consultants/FAO	Full (excl. mts.)	1971
Sierra Leone	Various	Part	1963-74
Solomon Islands	LRD	Full	1974
Somalia	Consultants	Limited	1969-1979
Sudan	Consultants	Part	1970-1979

Summary by countries

	Years of completion		
Extent of surveys	Largely pre-1970	Partly post-1970	Total
Full	20[1]	6	26
Part	4	3	7
Limited	0	2	2
Total	24	11	35

---

[1] West Indies other than Jamaica counted as one.

By 1970, twenty countries had completed full, systematic surveys. 'Full' means complete coverage of the country, 'systematic' that this was achieved by a national programme intended to achieve this. In most cases the work was conducted by national soil survey organizations; in the West Indies by the integrated Regional Research Institute. Exceptions are Brunei, surveyed by Huntings, Guyana by FAO, Papua New Guinea by CSIRO Australia, and Lesotho and the Seychelles through early work of the Land Resources Division. Government finance came mainly from the Colonial Office, although Sri Lanka benefited from Canadian support. New Zealand supported a number of Pacific island surveys.

The clearest case of a systematic survey by government is Uganda, where it was conceived and completed in a space of six years, 1956-1962. Ghana set out with the same intention, but the political problems which followed its early independence terminated the survey before completion. Malawi, with its modest one-man level of staffing, completed a national agro-ecological survey 1959-65 (although publication of the Southern Region was delayed). Malaya, having set up its soil classification through a joint project with New Zealand, conducted a series of regional surveys. The Regional Research Centre for the West Indies, with the willing support of governments, was able to survey most islands on a uniform basis. Sri Lanka and Tanzania achieved national coverage by means of compilation, synthesizing local surveys conducted over many years.

The next ten years, 1970-1980, saw a change in executing and financing agencies. British technical and financial support was primarily through the work of the Land Resources Division, which completed systematic surveys of Sabah and the Gambia, the latter an integrated agricultural development study. From 1966-1979 the Division also carried out land system surveys, coupled with land evaluation, covering a quarter of Nigeria. In Sudan, early work conducted by government was succeeded by large surveys by consultants for irrigation projects. FAO began work on Pakistan before it divided into West and East. In the West, it was able to build upon the very extensive surveys by consultants, mainly Huntings. In the East, now Bangladesh, FAO benefited from the sustained effort of Hugh Brammer in training local staff and guiding surveys; by the late 1970s this had led to it becoming the most comprehensively and completely surveyed country in the tropics. The last country to date to complete a full, systematic survey was Kenya, in which the local soil survey received prolonged financial, staff and training support from the Netherlands government.

By 1980, twenty-six countries were in possession of land resource surveys covering the whole of their territories, including soil or land system maps, together with some form of land evaluation or assessment of agricultural potential. Seven others, including Nigeria, Sudan and India, had partial although substantial coverage. The almost superhuman task facing the All-India Soil Survey continues to this day. Only two countries which left the Commonwealth, British Somalia (now part of Somalia) and Myanmar (formerly Burma) were left with very limited coverage.

In addition to surveys of national and regional extent, large numbers of local detailed studies were undertaken by national soil survey institutions. These were nearly always conducted in response to specific development requirements: the siting and internal layout of an experiment station, a commercial estate or smallholder development scheme, an irrigation project, and many other purposes. The outcome of most such surveys would be to go ahead with the proposed development — indeed, such decisions had often already been provisionally taken. Sometimes a survey would lead to modification of the project site by excluding unsuitable land. Less often, the recommendation would be to abandon it. Instances of the latter have already been noted: do not develop cocoa in Guyana, nor oil palm in the Gambia, and — too late — do not attempt mechanical clearance on the compact sandy soils in Tanzania.

There were many thousands of such local surveys. Most were not intended for attention outside the country concerned. Many were produced in mimeographed and stapled form, or sometimes manuscript copies in filing cabinets. For their immediate purposes such a non-permanent format did not matter, although it might mean that in years to come, good field survey work would be lost. Local surveys have always formed part of the output of national institutions, and now that broad reconnaissance is so far advanced, local studies make up much of their activity.

Applications: actual and potential uses of land resource surveys

The need for soil and land resource surveys is forcefully argued by soil scientists, but what has happened in practice? The potential uses were outlined in Chapter 1, under 'Why carry out soil surveys?' but the studies that have been made of their actual use are not reassuring. A review by Barry Dalal-Clayton and David Dent, *Surveys, Plans and People* (1993), concluded that planners and decision-makers ought to ask searching questions about natural resources, and do not do so. There is little coordination of activities between institutions:

> A range of initiatives is needed to develop land literacy and awareness of natural resources, the consequences of their mismanagement, and the value of natural resources information for environmental management and sustainable economic management.

It would take long experience of rural sector development in each country to produce a definitive answer on the extent to which survey information is put to use. In default of this, a source is the comments by correspondents on the use which has been made of their work.

Detailed surveys

The utility of detailed surveys offers no problem. A very large number of such surveys were made, all intended for specific and immediate local developments: to look into their desirability and to plan their layout. Many such developments went ahead: sugar, rubber, rice or pineapple production schemes, for example, each sited on suitable land. New agricultural experimental stations were set up, in the knowledge that the layout of trials must be adjusted to the distribution of soil types. A smaller number of proposed developments were abandoned on grounds that the land was found to be unsuitable. An unusual case was a proposal by the International Centre for Research in Agroforestry (ICRAF) to set up trials on the government's Muguga Research Station, where the soil was found to have such high inherent fertility that it would be of little use to study fertility improvement!

Robert Green writes, 'My personal surveys in Malawi were carried out in specific areas to satisfy an immediate requirement and were therefore all used.' For all such specific local developments, the preceding detailed surveys would never have been made if early implementation had not been intended.

Project surveys

The use made of project surveys is straightforward up to a certain stage, the point where the land use plan was completed and the decision to implement made. Some projects were found to be not viable, but many went ahead. Examples are major irrigation schemes in India and Pakistan, and land settlement schemes in Malaysia and in the Dry Zone of Sri Lanka. Not only did the Jengka Triangle project in Malaya go ahead, but in a follow-up study twenty years later, the map of the scheme bore a strong resemblance to the original soil and land suitability maps. The survey of Western Nigeria intended as a basis for replanting following swollen shoot disease in cocoa was in fact used for this purpose. In Malawi, what started as the British Irrigated Rice Project was changed in the course of the survey to a Sugar Project.

By and large, the existence of surveys meant that hazardous land was excluded from project development. In Ghana, Sandy Crosbie took part in work intended for expansion of areas under bananas, cocoa and pineapples. 'In some cases the benefits of such surveys were that they prevented money being spent on projects which were either not feasible or not economic.' The soils department of Booker Agriculture was set up because earlier they had set up a sugar estate on land which turned out to be unsuitable.

A basic and successful function was the allocation of land to different types of use and to specific crops. When it came to providing information on required management inputs and estimated outputs, however, the surveyors themselves often fell short of the ideal. For economic appraisal, what is required is estimates of crop yields (or other outputs) for specific soil types under one or more kinds of management. When those best qualified to make such estimates failed to provide them, the economists were forced to make their own assumptions, which were inevitably not specific to land types.

Another deficiency in making use of land resource information originates from the 'project' basis of development, the fact that the surveyors who make them are on contract terms. They complete the map, write their report, and then go home. Since they are not even present during the later stages of project planning, it is no wonder that land resource considerations receive less than adequate attention. Occasionally, soil surveyors have become project managers or directors, for example David Parry with Mott Macdonald and Stan Western with Huntings. The best way to remedy this situation is to ensure that staff from the local survey institution are included in the survey team, and that one of them takes part in the later stages of planning.

The potential for an ongoing role of soils information in monitoring and management after projects have been implemented is discussed below.

Reconnaissance surveys

With reconnaissance surveys, experience is variable. An extreme situation is that they were forgotten about. The most cynical view heard is that coloured soil maps make a fine decoration for the walls of government offices. Alun Jones reports an extreme case of non-use:

> Twenty years ago I sent a mission to Jamaica to build up a development project for World Bank financing. I told them to look for the soil survey maps we had compiled [the 'Green Books']. The mission returned to say that the Ministry of Agriculture were unaware that these existed. Such is the permanence of one's impact!

Other negative views are found. Graham Higgins, from experience in Northern Nigeria, considered that, 'The use of reconnaissance surveys has yet to be demonstrated.' Barry Dalal-Clayton writes of Zambia, 'My experience is that our reconnaissance work has never been used in strategic planning and policy making. Big decisions are generally taken on economic and political grounds; scientific data are rarely of significant influence.' David Eldridge writes, with reference to Botswana, 'Soil surveys are potentially very useful but it is difficult to get general-purpose maps used by government planners. Special-purpose surveys for individual projects are used much more, as the customer-client relationship is clear.'

In contrast, Charles Johnson, in sending me copies of Trapnell's surveys of Zambia (then Northern Rhodesia) remarked, 'They are somewhat the worse for wear, I'm afraid, having been much used during my 24 years as an Agricultural Officer there.' Anthony Mitchell writes that in surveys for rural development projects, both in Botswana and Malawi, he made considerable use of the earlier soil and agro-ecological surveys. 'These were enormously valuable as they brought together so much information.' Barry Dalal-

Clayton was recruited to survey an Integrated Rural Development Project in Tabora region, Tanzania, at a time when implementation had already begun:

> Not having learnt from the Groundnuts Scheme, they assumed they could just plough in with all the developments and that in due course the survey would confirm their wisdom. The project was eventually deemed a failure by Bank criteria.

John Bennett's comments on the use made of the land systems surveys of Nigeria, and my own experience with the agro-ecological survey of Malawi, have been noted above. A wide range of developments is detailed in Hugh Brammer's book, *Land Use and Land Use Planning* in Bangladesh (2002), for example, selecting sites for food-for-work schemes and for small-scale water development projects, scaling down of regional surveys for use in village agricultural development plans, zoning of land for urban development, flood protection and other hazard management.

FAO consultation for a project on Population-Supporting Capacity; at table from left, A. P. A. Vink (Netherlands), Jean King (UK, Unesco representative), Graham Higgins and Anthony Young (UK); behind: Maurice Purnell and Amir Kassam (UK), Rudi Dudal (Belgium). (*Graham Higgins*)

The use of surveys to site fertilizer trials and identify where the results should be applied is reported from several countries. In the Agro-Ecological Survey of Malawi the Agronomist, Peter Brown, made sure that the Soil Surveyor, myself, visited and identified the soil types on all the trial sites. Hugh Brammer stresses this aspect in Bangladesh, where the benefits from agricultural research were spread by the existence of an informal farmer-to-farmer extension process. In Northern Nigeria, Graham Higgins writes, 'An achievement of the early days was recognition of the coverage of representative (or in some cases not) experimental farms. To this day, experimental programmes are based on the areas covered by particular trials.' Gordon Anderson is remembered for his survey of the Kilimanjaro footslopes in Tanzania, but writes that, 'Survey took up only 20% of my time, the rest being dedicated to fertility studies and advice to agricultural, land planning and rangeland officers.' In the West Indies, Ian Twyford used soil survey results to set up a series of banana fertilizer experiments on the basis of soil types throughout the islands. 'Part of the work was to develop soil and leaf analytical standards and make them available to farmers.' Twyford then moved to a World Bank project in Pakistan and worked out fertilizer requirements on the various soils for the chief smallholder crops.

International and national analysis

A major use of reconnaissance surveys was to provide the soils input for international projects on developing countries. These began with the *FAO-UNESCO Soil Map of the World* at 1:5 000 000 scale, preparation of which was begun in 1961 and the ten sheets published 1970-80. Successive revisions followed, leading to the digitized version available today. Among FAO projects for which this provided an input were the determination of Agro-Ecological Zones, and the study of Potential Population-Carrying Capacities of Land of the Developing World. This last, based on populations of 1975, provided strikingly accurate forecast of food shortages in the year 2000.

National planning institutions in developing countries similarly used the results of soil surveys as an input for digitised processing of land resources, as geographical information systems. Early soil surveyors were not to know that their results, often little modified subsequently, would still be used in this way 50 or more years later.

Soil survey in developing countries today

Much of this review covers the period up to the end of the era of recon-naissance surveys, about 1975. It would not be complete, however, without a comment on the status of soil and land resource surveys in developing countries today.

For four countries covered in this book, Ghana, Sudan, Swaziland and Tanzania, the transition to post-independence surveys is described in the country-specific reviews of the World Soil Survey Archive and Catalogue (WOSSAC). By the end of 2016 there were nine of these: four countries covered in this book, Ghana, Sudan, Swaziland and Tanzania, together with the non-Commonwealth countries Angola, Iraq, Jordan, Liberia and Niger. Further country reviews are planned.

Tanzania is an example, a five-page review based on 545 archived maps and reports. The first section, The Colonial Surveys, outlines the main work recorded in this book: Milne's reconnaissance mapping and establishment of the catena concept; the work of Brian Anderson focused on soil fertility, and mapping by Gordon Anderson; C.G.T. Morison's concept of ethnopedology, farmers' knowledge of, and names for, their soils; and the failure of the East African Groundnuts Scheme, with the subsequent surveys which might have helped to avoid this investment disaster.

This is followed by a section on Post-independence Soil Mapping, conducted by, 'Large teams of expatriate and local surveyors.' Post-inde-pendence work has necessarily been largely financed externally. Mapping included detailed local surveys, directed at village-level planning and farm management-level documents. Thematic Assessments show how the experi-ence of the Groundnuts Scheme led to surveys directed at land suitability for specific crops: sisal, tea, navy beans, cocoa and cotton. 'Concern over environmental pressures and the supply of fuelwood' led to surveys of catchment forests in mountain areas. National Land Cover Mapping, based on satellite imagery and ground checking, was conducted by Huntings. Natural resources information was subsequently coordinated at the Univer-sity of Dar es Salaam, using data on water resources, vegetation changes, agro-economic zones and even population changes, to produce a Tanzanian

Natural Resources Information System (TANRIS), a digitised geographical information system.[1]

Soil institutions

Nearly all countries of any size have retained institutions with a mandate for land resource survey, development and management, although some have been weakened by reduced staffing. Most are either a section of the national agricultural research service, or are based on a university. FAO maintains a list of national soil information centres, mostly originating as soil surveys. Of the countries treated in this book, 24 are shown as having national soils institutions, half of these with more than one. Some retain the title of Soil Survey. Others expand their titles in the direction of applied soil science, such as the largest, the Indian Society of Soil Survey and Land Use Planning, the Bangladesh Soil Resource Development Institute, and the CSIR-Soil Research Institute of Ghana.

Soil science in Sri Lanka was conducted under the title of the Land Use Division, a section of the Irrigation Development Institute. Under the programme of work they list soil surveys at three levels of intensity: land suitability mapping, assessment of salinity and alkalinity, and surveys for proposed irrigation schemes and drainage projects. Malawi retains a soil survey section in the Department of Agriculture. The Malaysian soils service continues to contribute to land development and management.

The Kenyan Soil Survey, renamed the Land Resource Survey and Inventory, in conjunction with ISRIC, developed a Soil and Terrain Database (KENSOTER). This is essentially digitising, combining and processing pre-existing surveys of soil and other environmental factors. Other projects of interest are directed at soil monitoring. There is a Kenyan Soil Health Consortium. An assessment of Soil Organic Carbon Stocks and Change, was carried out 2002-05 in conjunction with the University of Reading and the Global Environment Facility. Other research projects are directed at soil degradation or soil health.

---

[1]   A retired soil surveyor, on being told that geographical information systems were now available, replied, 'Yes, we had those in my day. We called them maps.'

Ghana has a Soil Research Institute employing 18 scientists, situated at Kumasi, the former site of the Soil Survey. Among some 16 research projects are Promoting an Enabling Soil Health Policy Environment, Agricultural Climate Change Mitigation, and Soil Fertility Management Through Participatory On-farm Studies.

A positive factor is the wide range of techniques and services now at the disposal of surveys. Starting with the launch of the first Landsat satellite in 1972, multi-spectral scanning and other aircraft- and satellite-based types of remote sensing have offered new tools, particularly for reconnaissance surveys and for studies of vegetation and land use. Conventional air-photograph interpretation retains its primary value for more detailed work, and methods of field soil and vegetation survey have not changed greatly, but geographic information systems have opened up new possibilities for data processing, especially in the area of land evaluation. Spectral analysis for detecting soil degradation, and to some extent even identifying soil types, has been developed, for example by Keith Shepherd at the World Agroforestry Centre. Databases in climate, soils, irrigation, crop requirements, etc. are available from FAO and World Soil Information. The former problems of classification and correlation have been rationalised through the World Reference Base for Soil Resources.

Soils institutions in developing countries have qualified and dedicated professional staff, typically with a first degree from their own country followed by overseas postgraduate training. The fact that the institutions exist at all shows that governments continue to place value upon them. However, their operating budgets are frequently small, sometimes hardly sufficient to fund field transport. This means that they are unable to carry out substantial investigations without external aid.

Awareness and attitudes

A fundamental problem is a lack of awareness, by politicians, planners and developers, of the relevance of information on land resources. This came about in part through overcompensation for the former neglect of social aspects of rural development. Until the 1970s, studies of the day-to-day problems of farmers were either neglected or handled through farm system studies, the latter based on rather unimaginative methods of data collection.

As a result of project failures caused by neglect of the views of those who were being 'developed', the situation improved from the mid-1970s onward. A new approach was introduced, variously known as participatory rural appraisal, bottom-up planning, diagnosis and design, or more simply, talking with farmers. 'Stakeholders' became a much-used term, meaning the farmers and village institutions, regional and national governments, and donors anxious that aid should be efficiently spent.

The central scientific body for soil science, the International Union of Soil Sciences, declared 2015 to be the 'International Year of Soils'. Perhaps recognising that this had limited impact, it was followed by declaring 2015-24 the 'International Decade of Soils'. This, however, is preaching to the converted, and will have done little to raise the profile of soils within policy-making institutions and the community as a whole. Discussions centred on environmentalism rarely stress soils. The on-line encyclopedia, Wikipedia, writes that, 'The natural environment encompasses ... the interaction of all living species, climate, weather, and natural resources that affect human activity', listing the components of the environment as vegetation, microorganisms, soil, rocks, atmosphere, air, water and climate.

The UK government's Environment Agency is concerned mainly with water, flooding and waste. Its Department for International Development (DFID) is focused on poverty, the role of women, and climate change. When it comes to awareness and action on environment, soil usually takes a back seat.

Giving greater attention to social aspects was a positive and important change in attitudes. Unfortunately, it brought with it a neglect of the equally important role of natural resources and land management. A comparison of FAO publications illustrates the changing attitudes. In 1976 the *Framework for Land Evaluation* was almost entirely concerned with evaluation of resources, relegating 'economic and social appraisal' to a subsection. By 1993, *Guidelines for Land-Use Planning* contained a more balanced approach, setting 'Making the best use of limited resources' alongside 'Planning is for people'; there was still a substantial section on identifying opportunities for change, essentially land evaluation, but equal attention was given to involvement of the people affected by any such changes. Four years later in 1997, *Negotiating a Sustainable Future for Land* was very largely directed at social and institutional procedures, and balancing the interests of different stakeholders, with land resource aspects relegated to short sections; 'Soil (description, classification, mapping, suitability evaluation)' appears as one out of a list of thirty-three basic information needs — this despite its title referring to 'land'.

The pendulum had swung too far. David Parry, after forty years with consultant companies involved in project surveys, writes, 'When was the last time anybody carried out a soil and land capability study? Please do not tell me that all the necessary mapping has been done; I know for a fact that that is not the case.' The community of soil scientists is well aware of the low priority given to land resource aspects, and struggles to raise its profile. The *World Bank Development Report 2008* recognised that, 'GDP growth originating in agriculture is about four times more effective in raising incomes of extremely poor people ... than growth originating outside the sector', with soil being mentioned largely in the context of soil degradation.

One lever in this respect is the attention given to sustainability, a basic element of which is avoidance of land degradation. Another is the drive to reduce poverty, most of which is found in rural areas dependent on agriculture. A number of countries have research directed at assessing the implications of climatic change on soils and agriculture. There remains, however, a lack of awareness by national governments, international institutions and development agencies, that management of land resources plays a fundamental role in rural development. No amount of talking to farmers will get rid of phosphorus deficiency.

Achievements and lessons

Achievements

The achievements of land resource surveys were impressive. The reconnaissance surveys had set out the nature, extent and location of land resources over the greater part of Commonwealth countries. The knowledge gained formed the basis for the various applications of reconnaissance work: to estimate land development potential, provide a national framework of soil types, project identification and location, siting of agricultural research trials, and the potential to assist in agricultural extension. In short, the surveys of the period had found out 'what was there'.

A basic success of the post-war years was to establish that before any development involving land use change was attempted, a survey of land resources was essential. Project surveys fulfilled two main functions, avoidance of hazards, and provision of a rational basis for land use planning. These surveys, often carried out by consultant companies, became standard, particularly for irrigation projects where large initial investment was intended.

In addition, many local detailed surveys were completed during the Colonial period, and subsequently by post-independence national survey institutions. These are intended for local use, to confirm that sites intended for development are suitable, and to assist in their layout. Studies of this kind have a low profile outside the countries concerned, but they are numerous and form an important element in the work of survey institutions.

For how long do these surveys remain valid? Experience shows that their useful life is considerable, at least fifty years. The pioneering studies by Hardy in the West Indies and Milne in East Africa, completed without the essential aid of air-photograph interpretation, have been superseded, and Trapnell's ecological surveys of Zambia were incorporated into later soil studies. Surveys of the post-war period, however, remain in use. Updating of earlier work is possible but unlikely to be given priority. In broad terms, the nature and distribution of climate, geology, landforms and soils remain basically the same, whereas vegetation and land use have often radically changed.

Lessons

The first lesson learnt, early in the post-war period, was that natural resource surveys are an essential basis of rural development. The second, recognised some thirty years later, was that they are not sufficient. Studies of social aspects and institutions are equally necessary, with a key role given to discussions with farmers and others that have a stake in the use of land.

Regrettably, in appreciating the need for greater attention to the land users, international institutions, aid agencies and national governments have been in danger of losing sight of the equally important role of the land. FAO no longer has a staff member responsible for soil survey. The holdings of the World Soil Survey Archive and Catalogue (WOSSAC) show a sharp fall-off throughout the 1990s. When developments are proposed, it is all too common to find that no new land resource surveys will be carried out, with the statement that, 'We will make use of existing information.'

There has been insufficient recognition of the role of soils in three respects: monitoring, to record of changes in soil conditions, or soil health; addressing problems of farmers, as extension advice.; and integrating conservation with production, as conservation agriculture. These aspects are outlined further in Chapter 16.

A fundamental cause lies behind the neglect of soils by governments, the fact that natural resource aspects do not carry weight with development agencies. This is in some ways surprising, given the attention paid to environment and to sustainability.

Environment, however, is regarded mainly in terms of pollution. Conservation of land resources, which form the basis for rural livelihoods, does not receive the same attention. Sustainability was first conceived in terms of land. Sustainable land use is that which meets the needs of present land users, whilst conserving for future generations the resources on which that production depends — in short, sustainability equals production plus conservation. However, the concept was later extended to include institutional change, economic viability, and social acceptability. Important as these wider aspects are, they are apt to lead to a very vague definition of sustainability, which may lose sight of the essential need to conserve or improve land resources.

Land resource management and conservation is not an end in itself. Its role lies in contributing to the welfare of the rural population, present and future. In explaining the low priority given to survey and evaluation of land resources, and the part played by technical knowledge in improving their management, the compartmentalization of soil science must bear part of the blame. Soil scientists of today need to seek better recognition of the subject. This calls for wider horizons, an attempt to carry greater weight in development agencies and governments. There is an instructive case for comparison. Politicians and policy-makers have only to say the words, 'global warming' or 'climate change' and they are listened to. How to raise the profile of soil science, how to achieve better communication of its potential to improve human welfare, should be the priority for the scientists of today. Only in this way can the past achievements of land resource surveys fulfil their potential.

16

THE STUDY OF SOILS IN THE FIELD:
WHAT IS ITS ROLE TODAY?

In 1936 G. R. Clarke of Oxford University published his book *The Study of Soils in the Field*. This became a classic, running to five editions, the last, revised by his former student Philip Beckett, in 1971. Clearly the subject was in demand, meeting a need from students and professionals throughout the period when the field study of soils primarily meant soil survey.

This is no longer the case. The number of soil surveys being carried out has greatly decreased. The World Soil Survey Archive and Catalogue, based on the University of Cranfield has formed a collection of published surveys going back to the earliest times. Initially this was focused on British Commonwealth countries but it has now been extended to world coverage.

Holdings are currently in excess of 26 000. The graph shows acquisitions by date of publication. The pioneering period, predominantly the work of the individuals referred to in Chapter 2, ran from 1925-1941, typically with 10-30 maps and reports per year. Growth began during and immediately after World War II, the number of items passing 100 in 1950. From 1956-96 over 300 maps and reports per year were received. In this book the period 1950-1980 is referred to as the Golden Age of Soil Survey, and this was, indeed, the major period of national and regional reconnaissance surveys. However, activity continued at at a high level for a further 20 years, probably with a transition to more local surveys for development projects, particularly the voluminous output typical of consultancy reports.

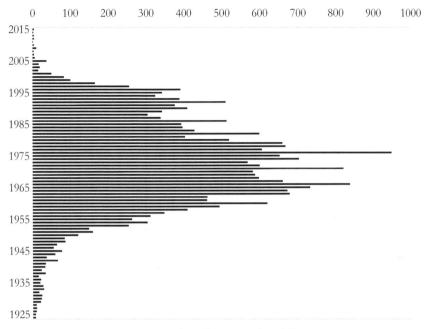

WOSSAC Holdings by Year of Publication

For ten years from 1995 there was a decline in activity, leading to a low level of items. There may be some element of time-lag in output reaching the archive but without doubt, rural development from 2006 onwards has frequently been carried out without systematic soil surveys. Why should this be so, and has it caused adverse effects? The negative answer would be that development agencies do not regard soil survey as necessary. The more positive approach is that the results from earlier work have proved to be of lasting value, supplying a foundation on which future development can be based.

An indication is provided by the changing titles of soils institutions. In Britain, the former Soil Survey of England and Wales first extended its title to become the Soil Survey and Land Research Centre, then in 2001 became the National Soil Resources Institute. Many developing country soil surveys have extended their titles in the direction of soil resource management or development, for example the Soil and Land-Use Survey of Ghana became the Soil Research Institute. It is clear that there remains a role for soil-based institutions but it is not primarily soil survey.

In World Bank projects it is standard to have a follow-up called Monitoring and Evaluation, but this largely focuses on economic results, and as far as I know has never included impacts upon the soil.

Extending the applications

Most developing countries retain soil research institutions, many having changed their titles to reflect their widening of interests and activities. There is still some survey activity, primarily where new irrigation projects are proposed, but research extends to support for land management, soil health, and crop-specific soil studies.

Publications in leading soils journals are now dominated by laboratory-based research., rightly directed at finding solutions to better soil management. There appears to be a danger, however, of losing sight of the complexity of soils in the field, and the variety of problems which arise from this. Reviewing the projects listed by developing country soil research institutions suggests that they may be better linked with field soil studies than their counterpart in the developed world. There are two areas in which the widening of perspective from survey offers opportunities: soil monitoring, and conservation agriculture.

Soil monitoring

In 1991 I published a paper in the journal *Soil Use and Management* entitled, 'Soil monitoring: a basic task for soil survey organisations.' At the time it was written there was very little to go on, the main example coming from Japan. Initially there was not much response, and the studies which began to appear were mainly concerned with soil pollution from industrial waste and acid rain. An FAO Conference on Soil Monitoring was held in 1994, by which time Hungary has set up observation points for observation of soil changes. Issues of *European Journal of Soil Science* in the first decade of the 21st century included a scattering of papers on soil monitoring in Europe. Linked with concern about climatic change, the role of soils as carbon stores received attention. There was a joint World Bank/FAO initiative attempting to define what were termed soil quality indicators. In the UK, a soil monitoring workshop was held at Rothamsted in 2008.

With respect developing countries of the tropics, the situation was transformed by a wider focus on land degradation, leading to the successive attempts at quantifying this on a world scale through the projects outlined in Chapter 13: on a world scale GLASOD and GLADA, and the Land Degradation in Drylands project (LADA) which included six country studies. There are now university courses in soil monitoring. The term 'soil health' has come into use, possibly better from a public awareness point of view, in that it is positive and covers land improvement as well as degradation.

Examples on a regional basis are the Soil Health Consortia for Eastern and Southern Africa, for example the Tanzania Soil Health Consortium, directed at ensuring that agriculture and soil management are on a sustainable basis. Technical advances both in both field observation and laboratory analysis have facilitated studies of soil health. As a method of remote sensing, reflectance spectroscopy from the air has had some success in identification of soils, particularly in highlighting areas of degradation. The low cost of spectral soil analysis in the laboratory makes it possible to deploy statistical sampling schemes over a target area, which allow estimates to be made of the prevalence of various soil constraints, including changes over time. Data of this kind can be combined high density ground observation.

At a national level, the former soil survey organizations should be engaged in the collection of data on the extent, location, degree and effects of land degradation. Action to check degradation, and to improve soil status, can only be taken if the consequences, for farmers and the national economy, are reliably and quantitatively estimated. By doing so, national soil research institutes could improve their status and contribute more to welfare.

When seeking advice on this topic, a sentence from a recent publication rang a bell in my mind:

> A basic task of national soil survey organizations should be to monitor changes in soil properties. The objectives would be to establish which land-use systems, on which soil types, are causing decline in fertility (Young, 1991).

> The scientific rigour of land health surveillance has the potential to provide a sound basis for directing and assess action to combat land

degradation. Specialized national surveillance units should be established … to provide integrated national land health systems. (Shepherd et al., 2015).

It is pleasing to learn that progress has been made.

The Conservation Agriculture Adviser

From the earliest days, one of the first specialists appointed by Departments of Agriculture was an Entomologist. He would be called out by District Agricultural Officers where it was believed that insects were damaging the crops. Sometimes this proved to be a microbiological problem, calling for advice from the Plant Pathologist. These specialists became a standard part of agricultural research and advisory centres. The main routine task involving the soil was to take samples for analysis in order to make fertilizer recommendations.

There could be comparable specialists on soil management. They would be a backup for the advisory service, called in when local problems, often of low crop yields, were believed to originate from the soil. Such problems might, indeed, arise through nutrient shortages for which the standard remedy was fertilization. But often it would be found that the soil had low organic matter status and biological activity, and poor physical conditions, hence giving a poor response to fertilizer. An example is the impermeable pans below the cultivated layer which became common in Malawi. Not only is the fertilizer solution often impractical on small farms in the tropics, but response might be uneconomic without measures being taken to restore soil physical conditions.

The concept of specialists in soil supporting the agricultural advisory service was suggested by my colleague at ICRAF, Peter Huxley, who suggested the title of Soil Technician. As this wording might be interpreted as referring to a sub-professional post, I subsequently modified it to Soil Management Adviser.

But why not make it Conservation Agriculture Adviser? This would reflect recent and ongoing developments in land husbandry. In the older approach, whilst Agricultural Extension staff were friends of the farmers, the soil conservation section sometimes appeared to be telling them to do things which had no short-term benefit. The approach and methods of conservation agriculture have transformed this situation. In effect, the former soil conservation staff would be integrated with the agricultural research service. Conservation Agriculture Units would include an element of adaptive research, directed at identified local problems, linked with backup to the extension service.

These are two ways in which the former focus on soil survey, a prime requirement for all types of soils research, has been changed into a more varied range of activities. There is also an ongoing need to link trials of crop requirements and management to the landscapes on which farmers must apply them. There is a sequence of types of soils research: basic, applied, and adaptive, broadly corresponding to laboratory studies, field station trials, and on-farm research. The study of soils in the field brings benefits of two kinds: to indicate the problems which research needs to address, and to convert the results to practical farming recommendations. In such ways, the legacy of early soil surveys can be applied to ongoing work directed at conservation and improvement of the land resource base on which agriculture depends.

The integral nature of population policy

What has population policy got to do with land resource management, rural development, reduction of hunger and poverty, and land degradation? The view most often found today is that population policy is a separate issue. Population increase in the developing world is treated as 'given'. This is the attitude found in pronouncements by most of the major international institutions, including the World Bank, FAO, and the International Food Policy Research Institute (IFPRI). They quote estimates of future populations, usually up to 2050, and treat them as external, independent, variables. From the 1960s onwards a neo-Malthusian view arose, that the world is already too highly populated to sustain its limited resources. This view, however, does not command attention from most UN institutions, nor from policy-makers at large.

There are exceptions. Some academic economists now argue that population increase in developing countries has a negative impact on economic growth, poverty, equality, and sustainable use of natural resources.

Conversely, a decline in the rate of population growth fertility makes an appreciable contribution to reducing poverty. This applies most clearly to countries with a large rural sector and high dependence on agriculture, the low-income food-deficit countries (LIFDCs). In non-jargon terms, these are the world's poorest countries. A view has been expressed that the 'green revolution' can be considered a failure, in that the large increases in crop yields has been nullified by growth of populations which they support.

A striking early view was expressed by the Director of Agriculture in Malawi, R. W. Kettlewell, over 60 years ago: 'Progress [in agricultural development] will be nullified unless Nyasaland's present rate of growth of population is substantially reduced.' At the time he wrote this, 1955, his country's population was 3.25 million. By 2017 it had reached 18 million and was still rising at 3% a year. Average farm size is less the half a hectare.

Only one major international institution campaigns for the need for greater action, on population change. This is the UN Fund for Population Activities (UNFPA), which forcibly sets out arguments for the adverse effects of rapid population increase. It is also found in a major non-governmental organization, People and Planet. A Royal Society review meeting in 1998 had the self-explanatory title, 'Land Resources: on the Edge of the Malthusian precipice?' Most significantly, a meeting of scientific academies from around the world, both developed and developing countries, issued this statement in 1993:

> There is no doubt that the threat to the [global] ecosystem is linked to population size and resource use ... Family planning could bring more benefits to more people at less cost than any other single technology ... Success in dealing with global social, economic and environmental problems cannot be achieved without a stable world population ... We must achieve zero population growth within the lifetime of our children.

Government attempts to check population increase by coercive means failed in India, and succeeded only under the autocratic regime of China, using methods not acceptable to the world at large. There is no need to use this approach. At the UN International Conference on Population and Development in 1994, a consensus was reached on an ethically acceptable package for reducing rates of population growth. There were three elements to this:

improvements in the education and status of women; improvements in reproductive health, both of mothers and children; and provision of access to family planning services for all.

Some progress is being made. The percentage increase in world population, over 2% from 1962 to 1971, has fallen to close to 1%. However, as this is on an ever-larger base the annual increase, which reached a maximum of 88 million in 1989, since 2006 has remained close to 78 million. To put it in simplified terms, every day some 360 000 people are born and 150 000 die. Nearly all the increase is in developing countries, which places a colossal ongoing burden on governments intent on reducing poverty and hunger. Famines in Africa recur at frequent intervals, often triggered by civil unrest or similar political causes, but with the suffering becoming greater each time as larger populations are affected.

Statements on population policy have mostly been made by economists and demographers. Approaching the question from a quite different viewpoint, as a land resource scientist, I am inexorably led to the view that efforts to reduce poverty and hunger will constantly be thwarted, sometimes nullified, by population increase. Sustainable use of land resources, avoiding degradation, will not be achieved. This conclusion is based partly on statistical sources, but more generally from travel and observation in the field. The key prerequisite for action is awareness, by international institutions and national governments, of the link between population increase and welfare. Greater awareness would bring about a change in attitudes, which in turn would lead to action.

It is a long way from soils to population, yet I am led to the conclusion that population policy is an integral part of agricultural development. Others have reached the same conclusion, notably the hydrologist and former Deputy Director of the East African Agriculture and Forestry Research Organisation Sir Charles Pereira, who devoted his energies during retirement to the problem of population increase. But if such an influential voice as his, and that of the naturalist David Attenborough, cannot bring about such a change, what hope have the rest of us got?

Checking population increase is by no means a panacea. It needs to be combined with the known methods of agricultural research and development. When putting forward this view in *Geographical Journal* 2005, I ended with a passage which in some respects is the conclusion of a career which started in soil survey:

> Emotion is normally out of place in scientific statements, but the statistics and rhetoric of internal discussion can be translated into individual suffering. After a lifetime of seeking to promote the welfare of peoples of the tropics through work on the survey, evaluation and management of land resources, one feels deeply for the plight of families, caught in a state of endemic poverty and hunger through circumstances beyond their control. Without more explicit recognition that action to check population increase is an integral element in policy on poverty, food security, and sustainability, this suffering will continue.

BIBLIOGRAPHY

The following bibliography contains maps and reports referred to in the text. Most can be consulted at the World Soil Survey Archive and Catalogue (WOSSAC), Cranfield University, UK (www.wossac.com), ISRIC--World Soils Information (www.isric.org), or the European Soil Data Centre (ESDAC) (www.esdac.jrc.ec.europa.eu).

The *Tropical Agriculture Association Newsletter* is the forerunner of *Agriculture for Development*, both produced by the Tropical Agriculture Association (UK).

Much of the biographical and anecdotal material is derived from personal communications, acknowledgement and warm thanks for which are given on p. xxi.

Chapter 1 Setting the task

Introduction

Brevik, E. (ed.) 2006 *History of soil science in developing countries*. Abstracts, 18th World Congress of Soil Science, Philadelphia, USA. CD-ROM. See Section 4.5A, Sessions 38 (5 oral papers) and 173 (16 poster papers). scisoc.confex.com/crops/wc2006/techprogram/S2120.HTM

D'Hoore, J. L. 1964 *Soil map of Africa scale 1 to 5 000 000: explanatory monograph.* (Map in separate folder.) Commission for Technical Co-operation in Africa Publication 93. Lagos: CCTA.

FAO-UNESCO 1970-80 *Soil map of the world 1:5 000 000*. Volumes 1-10. Paris: UNESCO.

Shantz, H. L. and Marbut, C. F. 1923 *The vegetation and soils of Africa*. American Geographical Society Research Series 13. New York: National Research Council.

Soil science in the United Kingdom

Clarke, G. R. 1936 *The study of soils in the field*. London: Oxford University Press. Five editions, the last revised by P. H. T. Beckett, 1971. (The first edition was entitled *The study of the soil in the field*.)

Darwin, C. 1881 *The formation of vegetable mould, through the action of earthworms*. London: Murray.

Defries, A. 1938 *Sheep and turnips: being the life and times of Arthur Young, F.R.S., First Secretary to the Board of Agriculture*. London: Methuen.

Hall, A. D. and Russell, E. J. 1911 *A report on the agriculture and soils of Kent, Surrey, and Sussex*. London: HMSO.

Russell, E. J. 1912 *Soil conditions and plant growth*. [Rothamsted] Monographs on Biochemistry. London: Longmans Green. Editions 2-10 by E. J. Russell and E. W. Russell. Eleventh edition 1988, as *Russell's soil conditions and plant growth*, edited by A. Wild. Twelfth edition 2013, with the original title, edited by P. J. Gregory and S. Nortcliff.

Russell, E. J. 1966 *A history of agricultural science in Great Britain 1620-1954*. London: Allen and Unwin.

Young, Arthur 1804 *General view of the agriculture of the County of Norfolk*. London: Board of Agriculture. (One of a series of 'General Views' for counties of the United Kingdom by Arthur Young and colleagues, published 1793-1813.)

Soil science around the world

Brevik, E. (ed.) 2006 See Introduction above.

Helms, D. 2006 Soil survey in Puerto Rico: a brief history. *In History of soil science in developing countries* (ed. E. Brevik). See Introduction above.

McDonald, P. 1994 *The literature of soil science*. Ithaca, USA: Cornell University Press.

Mohr, E. C. J. and van Baren, F. A. 1954 *Tropical soils*. The Hague, Netherlands: van Hoeve.

Vageler, P. 1933 *Grundriss der tropishcen und subtropischen Bodenkunde*. Berlin. Translated by H. Greene 1933 as *Introduction to tropical soils*. London: Macmillan.

Warkentin, B.P. (ed.) 2006 *Footsteps in the soil: people and ideas in soil history*. Amsterdam: Elsevier.

Yaalon, D. H. and Berkowicz, S. (eds.) 1997 History of soil science: international perspectives. *Advances in Geoecology* 29: 1-438.

Chapter 2 The pioneers

Explorers and scientific travellers

Buchanan, F. 1807. *A journey from Madras through the countries of Mysore, Canara and Malabar*. 3 vols. London: East India Company.

Johnston, H. 1902 *The Uganda Protectorate*. 2 vols. London: Hutchinson.

Livingstone, D. and Waller, H. 1874 *The last journals of David Livingstone*. London: Murray. 2 vols.

Low, Capt. J. 1836 *Soils and agriculture of Penang*.

Lugard, Baron F. D. of Abinger 1893 *Map shewing routes of Capt. Lugard in Uganda and Unyoro and adjoining territories with notes on soil, products, vegetation, etc*. Map scale 1:506 880.

Early scientific accounts

Bruce, A. 1923 *Forest soils of Ceylon*. Department of Agriculture Bulletin 61. Colombo.

Castens, H. E. 1927 *An investigation of the soil conditions in Compartment 1, Bwet Reserve, Prome Division, with reference to the dying off of Tectona grandis. Burma Forest Bulletin 18*. Rangoon.

Charlton, J. 1932 *Soil survey – Mandalay Canal area, Burma*. Agricultural Surveys 15. Rangoon: Department of Agriculture.

Charlton, J. 1935 *The sugarcane soils of Burma*. Agricultural Surveys 22. Rangoon: Department of Agriculture.

Gracie, D. S. 1930 *Notes on a preliminary soil survey of some parts of Kenya Colony*. Kenya Agricultural Bulletin 7. Nairobi: Government Printer.

Hughes, J. 1879 *Ceylon coffee soils and manures. A report to the Ceylon Coffee Planters Association*. London: Straker.

Kirkham, V. H. 1913 *Kenya: soil survey: hints on collecting samples*. Agricultural Leaflet 5. Nairobi: Government Printer.

Lefroy, J. H. 1873 *Remarks on the chemical analyses of samples of soils from Bermuda*. Hamilton, Bermuda: Bermudian Office.

Warth, F. J. 1916 *Note on the soil of experimental farms*. Department of Agriculture Bulletin 13. Rangoon.

Wright, C. H. 1916 *Fiji: report on the soils*. Fiji Agricultural Bulletin 9.

Frederick Hardy

N. Ahmad 1977 In memoriam: Emeritus Professor F. Hardy – Honorary Member I.S.S.S. (1889-1977). *Bulletin of the International Society of Soil Science* 52: 30-31.

Hardy, F. 1929 Some aspects of soil survey work. *Journal of Agricultural Science* 19: 1-18.

Hardy, F. 1930 Tropical soil surveys. Editorial and Parts I-III. *Tropical Agriculture* 7: 229-31, 235-36, 274-77, 300-1.

Hardy, F., Rodrigues, G. and 3 others, 1922-1936 and 1947. *Studies in West Indian soils. I-XIII.* Port of Spain, Trinidad: Government Printer. (Nos. I and II in *West Indies Bulletin* Vol. 19 (1922); Nos. III-XIII issued as supplements to *Tropical Agriculture.* (The 'Grey Books'.)

E. W. Russell 1977 Fred Hardy [obituary]. *Nature* 269: 93-94.

Willcox, O. W. 1929 Review of O. Arrhenius, Soil Survey in Java. *Facts about Sugar* 8 September 1928; with ensuing discussion by F. S. Earle, V. J. Koningsburger, and A. G. McCall, *ibid.* 13 October and 8 December 1928, 19 January and 16 February 1929.

Arthur Hornby

Dixey, F., Clements, J. R., and Hornby, A. J. W. 1924 *The destruction of vegetation and its relation to climate, water supply, and soil fertility.* Department of Agriculture Bulletin 1 of 1924. Zomba: Government Printer.

Hornby, A. J. W. 1925 *Soil survey of Nyasaland: high-level lateritic soils.* Department of Agriculture Bulletin 1 of 1925. Zomba: Government Printer.

Hornby, A. J. W. 1934 *Denudation and soil erosion in Nyasaland.* Department of Agriculture Bulletin 11 (New Series). Zomba: Government Printer.

Hornby, A. J. W. 1938a *Report of the agricultural survey of the five most northerly districts of Nyasaland*, 103 pp. Zomba: Government Printer.

Hornby, A. J. W. 1938b *Soil map of central Nyasaland.* Zomba: Government Printer. Intended to accompany Hornby (1938a) but published separately.

Hornby, A. J. W. and Maxwell, W. A. 1935 *Summary of the results of the agricultural survey of Central Nyasaland.* Mimeographed, 102 pp., with printed Map showing Agricultural Divisions. Zomba: Government Printer.

Frederick Martin

Martin, F. J. 1926 *Report on the survey of the soils of the Colony and Protectorate of Sierra Leone.* Freetown: Government Printing Office.

Martin, F. J. and Doyne, H. C. 1932 *Soil survey of Sierra Leone*. Freetown: Government Printer.

Geoffrey Milne

Anon. 1942 Milne, G. [obituary]. *Nature* 149: 188.

Milne, G. 1935 Composite units for the mapping of complex soil associations. *Transactions of the 3rd International Congress of Soil Science* Vol. 1: 345-47.

Milne, G. 1935 Some suggested units of classification and mapping, particularly for East African soils. *Proceedings of the 2nd Conference of East African Agricultural Chemistry*, 1934, pp. 48-49. *Soil Research* 4, (1935), 183-198.

Milne, G. 1936 Normal erosion as a factor in soil profile development. *Nature* 138: 548-49.

Milne G. 1936 *A provisional soil map of East Africa (Kenya, Uganda, Tanganyika and Zanzibar) with explanatory memoir*. Amani Memoirs. Amani, Tanganyika. Map at 1:2 000 000 scale and 34 pp.

Milne, G. 1936/1947 A soil reconnaissance journey through parts of Tanganyika Territory, December 1935 to February 1936. *Journal of Ecology* 35 (1947): 192-265.

Milne, K. 1978 Geoffrey Milne 1898-1942 [Obituary]. In *Bibliographical studies of geographers* (ed. T. W. Freeman and P. Pinchemel), Vol. 2, 89-92.

Colin Trapnell

Brunt, M. 2004 Colin Trapnell OBE, 1907-2004. *Tropical Agriculture Association Newsletter* June 2004, 29.

[Brunt, M.] 2004 Colin Trapnell [obituary]. *The Times*, 20 February 2004.

Burtt, B. D. (ed. C. H. N. Jackson) 1942 Some East African vegetation communities. Burtt Memorial Supplement. *Journal of Ecology* 30: 65-146.

Oxford University Exploration Club 2005 *Colin Trapnell OBE (1907-2004) [obituary]*.

Trapnell, C. G. 1943 *The soils, vegetation and agriculture of North-Eastern Rhodesia: report of the ecological survey*. Lusaka: Government Printer.

Trapnell, C. G. and Brunt, M. 1987 *Vegetation and climate maps of S. W. Kenya*. Memoir and eight maps. Tolworth, London: Land Resources Development Centre.

Trapnell, C. G. and Clothier, J. N. 1937 *The soils, vegetation and agricultural systems of North Western Rhodesia: report of the Ecological Survey*. Lusaka: Government Printer. A *Provisional vegetation-soil map at scale 1:2 000 000* is bound into this report.

Trapnell, C. G., and Smith, P. 2001 *Ecological survey of Zambia: the traverse records of C. G. Trapnell* 1932-1943. Vols 1-3. Kew, London: Royal Botanic Gardens. Volume 3 is a reprint of the map in Trapnell et al. (1950).

Trapnell, C. G., Martin, J. D. and Allan, W. 1950 *Vegetation-soil map of Northern Rhodesia*. Memoir with map in two sheets, scale 1:1 000 000. Map reprinted as Vol. 3 of Trapnell and Smith (2001).

Chapter 3 Preparing the ground

Attitudes and objectives

Hailey, Lord 1938 *An African survey*. London: Oxford University Press.

Institutions: the Colonial Service

Kirk-Greene, A. 1999 *On crown service: a history of HM Colonial and Overseas Civil Services*. London: Radcliffe.

Kirk-Greene, A. 2005 So few in the field: the staffing of the Colonial Agricultural, Forest and Veterinary services. In *How green was our empire? Environment, development and the Colonial Service* (ed. T. Barringer, Institute of Commonwealth Studies, London), 23-30.

Masefield, G. B. 1972 *A history of the Colonial agricultural service*. Oxford, UK: Clarendon.

The Land Resources Division

Baulkwill, W. J. 1972 The Land Resources Division of the Overseas Development Administration. *Tropical Science* 14: 305-22.

Brunt, M. 2000 Michael Bawden, David Pratt, and the Land Resources Development Centre. *Tropical Agricultural Association Newsletter* 20(3): 27-31.

Makin, J., Bennett, J., Brunt, M., and Griffin, C. (eds.) 2006 *Developing countries. Evaluation of land potential. The work of LRD, 1956-2001*. Bucknell, UK: LRD Publishers.

Consultant companies

Landon, R. (ed.) 1984 *Booker tropical soil manual*. London: Longman.

Robertson, V. n.d. *HTS – a personal history*. Mimeo (39 pp.). Hemel Hempstead, UK: HTSPE.

Western, S. 1978 *Soil survey contracts and quality control*. Oxford, UK: Clarendon.

Methods of survey: air-photograph interpretation

Bourne, R. 1928 *Aerial survey in relation to economic development of new countries, with special reference to an investigation carried out in N. Rhodesia*. Oxford Forestry Memoir 9.

Hotine, M. 1927 *Simple methods of surveying from air photographs*. London: HMSO.

Kemp, R. C. et al. 1925 *Aero-photo survey and mapping of the forests of the Irrawaddy Delta, Burma*. Forestry Bulletin 11.

National Collection of Aerial Photography, Edinburgh.
www.ncap.org.uk

The land systems approach

Brink, A. B., Mabbutt, J. A., Webster, R. and Beckett, P. H. T. 1966 *Report of the working group on land classification and data storage*. MEXE Report 940. Christchurch, UK: MEXE.

Christian, C. S. and Stewart, C. A. 1953 *Survey of the Katherine-Darwin region, 1946*. Land Research Series 1. Melbourne: CSIRO.

Tropical soil science

Bramao, D. L. 1968 *The first draft soil map of the world*. Reprint Bulletin 34, International Society of Soil Science.

Charter, C. F. 1957 *Suggestions for a classification of tropical soils*. Miscellaneous Paper 4. Kumasi, Ghana: Division of Soil and Land Use Survey.

d'Hoore, J. L. 1964 *Soil map of Africa scale 1:5 000 000: explanatory monograph*. CCTA Publication 93. Lagos: CCTA.

Dudal, R. 1968 *Definitions of soil units for the Soil Map of the World*. World Soil Resources Reports 33. Rome: FAO.

FAO 2015 *World reference base for soil resources 2014: a framework for international classification, correlation and communication*. Rome: FAO. (Earlier versions from 1994)

FAO-UNESCO 1970-80 *Soil map of the world 1:5 000 000*. Volumes 1-10. Paris: UNESCO.

Soil Survey staff 1960 *Soil classification: a comprehensive system: 7th approximation*. Washington DC: US Department of Agriculture.

Soil Survey staff 1975 *Soil taxonomy: a basic system of soil classification for making and interpreting soil surveys*. Agriculture Handbook 426. Washington DC: US Department of Agriculture. Second revised edition 1999.

Carrying out a soil survey

Dent, D. and Young, A. 1981 *Soil survey and land evaluation*. London: Allen and Unwin.

FAO 1977 *Guidelines for soil profile description*. Rome: FAO.

Soil Survey Staff 1951 *Soil survey manual*. Agriculture Handbook 18. Washington DC: US Department of Agriculture.

Young, A. 1968 Natural resource surveys for land development in the tropics. *Geography* 53: 229-248.

Young, A. 1973 Soil survey procedures in land development planning. *Geographical Journal* 139: 53-64.

Young, A. 1976 *Tropical soils and soil survey*. Cambridge, UK: University Press.

Young, A. 1998 *Land resources: now and for the future*. Cambridge, UK: University Press.

Interpreting surveys: land evaluation and land use planning

FAO 1976 *A framework for land evaluation*. Soils Bulletin 32. Rome: FAO.

FAO 1993 *Guidelines for land-use planning*. Development Series 1. Rome: FAO.

FAO-UNEP 1997 *Negotiating a sustainable future for land*. Rome: FAO and UNEP.

Klingebiel, A. A. and Montgomery, P. H. 1961 *Land capability classification*. Agriculture Handbook 210. Washington DC: US Department of Agriculture.

US Bureau of Reclamation 1953 *Bureau of Reclamation Manual. Vol. V, Irrigated land use. Part 2, Land classification*. Washington DC: US Bureau of Reclamation.

Chapter 4 East Africa

Uganda

Johnston, H. 1902 The Uganda Protectorate. London: Hutchinson, two volumes.

Langdale-Brown, I., Osmaston, H. A., and Wilson, J. G. 1964 The vegetation of Uganda and its bearing on land-use. Entebbe: Government of Uganda.

Memoirs of the Research Division, Department of Agriculture. Series 1, Soils; Series 2, Vegetation; Series 3, The systems of agriculture practised. Each Series as Volumes 1-6, in part consolidated:

Soils: E. M. Chenery, C. D. Ollier, R. Radwanski, J. G. Wilson, 1959-60.

Vegetation: I. Langdale-Brown and G. Wilson, 1960-62.

The systems of agriculture practised, D. J. Parsons, 1960.

Tothill, J. D. (ed.) 1938 *A report on nineteen surveys done in small agricultural areas in Uganda with a view to ascertaining the position with regard to soil deterioration.* Entebbe: Government Printer.

Uganda Government 1967 *Soils. Map, scale 1:1 500 000.*

Kenya

International Union of Soil Sciences 2003 *In memoriam: Wim Sombroek 1934-2003.* www.iuss.org/index.php?article_=92

Kenya Surveys Department 1959 *National atlas of Kenya.* Nairobi. 2nd Edition 1962, 3rd Edition 1970.

Kirkham, V. H. 1913 *Kenya: soil survey: hints on collecting samples.* Agricultural Leaflet 5. Nairobi: Government Printer.

Scott, R. M., Webster, R. and Lawrence, C. J. 1971 *A land system atlas of Western Kenya.* Military Engineering Experimental Establishment (MEXE), Christchurch, UK.

Sombroek, W. G., Braun, H. M. H., and van der Pouw, B. J. A. 1980 *Exploratory soil map and agro-climatic zone map of Kenya, 1980, scale 1:1 000 000.* Exploratory Soil Survey Report El. Nairobi: Kenya Soil Survey.

Trapnell, C. G. and Brunt, M. 1987 *Vegetation and climate maps of S. W. Kenya.* Memoir and eight maps. Tolworth, London: Land Resources Development Centre.

Tanzania (Tanganyika and Zanzibar)

Anderson, B. 1957 *A survey of the soils of the Kongwa and Nachingwea Districts of Tanganyika.* Dar es Salaam: Tanganyika Agricultural Corporation.

Anderson, G. D. 1961 *Soil survey of Maramba Estate with particular view to the cocoa potential.* Arusha, Tanganyika: Northern Research Centre.

Anderson, G. D. 1965 Soil factors affecting the distribution of the grassland types and their utilization by wild animals on the Serenged Plains, Tanganyika. *Journal of Ecology* 53: 33-56.

Anderson, G. D. 1968 *A survey of the soils and land use potential of the southern and eastern slopes of Mount Kilimanjaro, Tanzania.* Arusha, Tanganyika: Northern

Research Centre. Reprinted 1975 by Bureau of Resource Assessment and Land Use Planning (BRALUP), University of Dar es Salaam.

Anderson, G. D. 1973 *A survey of the soils and land use potential of the Lokisale and Galappo areas of northern Tanzania, particularly with a view to Navy Bean production.* Oxford, UK: FAO/IBRD Cooperative Programme

Anderson, G. D. 1998 Grow the soils to grow the crops in Africa. In *Soil quality and agricultural sustainability* (ed. R. Lal, Ann Arbor Press, Chelsea, Michigan), 237-50.

Anderson, G. D. et al. 1973 Soil factors affecting the distribution of the vegetation types and their utilization by wild animals in Ngorongoro Crater, Tanganyika. *Journal of Ecology* 61: 627-51.

Charter, C. F. 1950/58 *Report on the environmental conditions prevailing in Block 'A', Southern Province, Tanganyika Territory. Accra, Ghana: Department of Agriculture.* (Original report in typescript, 1950; published, 1958.)

Crompton, E. 1953 Grow the soil to grow the grass. *Agriculture* 60: 301-10.

Government of Tanganyika 1967 *National atlas of Tanganyika.* Dar es Salaam, Government of Tanganyika. Includes map of Soils [G. D. Anderson].

Fernandes, E. C. M., O'Kdng'ati, A., and Maghembe, J. 1984 The Chagga home-gardens: a multistoried agroforestry cropping system on Mount Kilimanjaro. *Agroforestry Systems* 2: 73-86.

Rapp, A., Berry, L. and Temple, P. H. (eds.) 1972 Studies of soil erosion and sedimentation in Tanzania. *Geografiska Annaler* 54A: 105-379.

Woomer, P. L. and Swift, M. J. (eds.) 1994 *The biological management of soil fertility.* Chichester, UK: Wiley.

WOSSAC, n.d. *Overview of Tanzanian materials.* www.wossac.com/archive/overview_tanzania.cfm

Zanzibar

Calton, W. E., Tidbury, G. E., and Walker, G. F. 1955 A study of the more important soils of Zanzibar Protectorate. *East African Agricultural Journal* 21: 53-60.

Sudan

Barnett, T. 1977 *The Gezira Scheme: an illusion of development.* London: Cass.

Gaitskill, A. 1959 *Gezira: a story of development in the Sudan.* London: Faber.

Tothill, J. D. (ed.) 1948 *Agriculture in the Sudan*. London: Oxford UP. (Chapters 7, Origins of soils, by J. D. Tothill; 8, Soils, by H. Greene; and 20, Review of experimental work, by F. Crowther.)

WOSSAC, n.d. *The historical context for land resource surveys in North and South Sudan*. www.wossac.com/archive/overview_sudan.cfm

Herbert Greene

Anonymous 1964 Dr Herbert Greene [Obituary]. *The Times* 1 May, p. 17.

Greene, H. 1928 Soil profile in the eastern Gezira. *Journal of Agricultural Science* 18: 518-530.

Greene, H. 1961 Some recent work on soils of the humid tropics. *Soils and Fertilizers* 24: 325-27.

Greene, H. 1963 Prospects in soil science (Presidential Address). *Journal of Soil Science* 14: 1-11.

Chapter 5 West Africa

Ghana (Gold Coast)

Ahn, P. 1961 *Soils of the Lower Tano Basin*. Memoir 2, Soil and Land Use Survey. Kumasi, Ghana: Ministry of Food and Agriculture.

Brammer, H. 1967 *Soils of the Accra Plains*. Memoir 3, Soil and Land Use Survey. Kumasi, Ghana: Ministry of Food and Agriculture.

Nye, P. H. and Greenland, D. J. 1960 *The soil under shifting cultivation*. Commonwealth Bureau of Soils, Technical Communication 51. Famham Royal, UK: Commonwealth Agricultural Bureaux.

Obeng, Henry Benjamin: biography. www.ghanaweb.com/GhanaHomePage/people/person.php?ID=206

Radwanski, S. A. unpublished. *Upper Tano Basin: final report: draft*. Typescript, with 19 hand-drawn maps at 1:250 000. Now in the WOSSAC collection.

Reports of the Soil and Land Use Survey Division, for 1951-55, 1956, 1957. Kumasi, Ghana: Ministry of Agriculture.

Wills, J. B. (ed.) 1962 *Agriculture and land use in Ghana*. London: Oxford University Press. Nigeria.

WOSSAC n.d. *WOSSAC material held in the Archive for Ghana*. www.wossac.com/archive/overview_ghana.cfm

Nigeria

Land Resources Development Centre, Tolworth, UK, 1966-79. Land Resource Studies (LRS) and Land Resource Reports (LRR):
The land resources of Southern Sardauna and Southern Adamawa Provinces, Northern Nigeria. LRS 2, 1966.
Land resources of North East Nigeria. LRS 9, 1972.
Soils of the Western State savanna in Nigeria. LRS 23, 1976.
Land resources of Central Nigeria: environmental aspects of the Kaduna Plains. LRR 19, 1977. (Supplementary to LRS 29)
Land resources of Central Nigeria. LRS 29, 1979.

Smyth, A. J. and Montgomery, R. F. 1962 *Soils and land use in central Western Nigeria.* Ibadan: Government of Western Nigeria.

Harry Vine

Vine, H. 1953 Experiments on the maintenance of soil fertility at Ibadan, Nigeria. *Empire Journal of Experimental Agriculture* 21: 65-85.

Vine, H. 1966 Tropical soils, in *Agriculture in the tropics* (ed. C. C. Webster and P. N. Wilson, Longmans, London).

Vine, P. 2005 Dr Harry Vine [obituary]. *IUSS Bulletin* 107.

Vine, P. 2005 Dr Harry Vine [obituary]. *British Society of Soil Science Newsletter* 47, July 2005, 21.

Sierra Leone and Gambia

Birchall, C. J. et al. 1963 Land suitability evaluation in Sierra Leone: methodology and results. In *National Seminar on Land and Water Resources Survey (ed. D. C. Schwaar). Typewritten. Freetown: Department of Agriculture.*

Dunsmore, J. R. et al. 1976 *The agricultural development of The Gambia: an agricultural, environmental and socioeconomic analysis.* Land Resource Study 22. Tolworth, UK: Land Resources Division.

Hill, I. D. 1969 *An assessment of the possibilities of oil palm cultivation in Western Division, The Gambia.* Land Resource Study 6. Tolworth, UK: Land Resources Division.

Odell, R. T. et al. 1974 *Characteristics, classification and adaptation of soils in selected areas in Sierra Leone, West Africa.* Bulletin 748, College of Agriculture, University of Illinois at Urbana-Champaign, USA.

Sivarajasingham, S. 1968 *Soil and land use survey in the Eastern Province.* Report to the Government of Sierra Leone. Rome: FAO.

Stobbs, A. 1963 *The soils and geography of the Boliland region of Sierra Leone*. Freetown: Government of Sierra Leone.

Chapter 6 Biographical interlude

Cecil Frederick Charter

Brammer, H. 1960 *Cecil Frederick Charter [Obituary]*. Annual Report of the Soil Survey Department, Gold Coast, 1960. Includes list of Charter's publications.

Charter, C. F. 1937 *Soil survey (reconnaissance) of Antigua and Barbuda, Leeward Elands*. Antigua, Government Secretariat.

Charter, C. F. 1948 *Cocoa soils: good and bad. An introduction to the soils of the forest regions of West Africa*. Tafo, Gold Coast: West African cocoa Research Institute (mimeo).

Charter, C. F. 1949 *Methods of soil survey in use in the Gold Coast*. Bulletin Agricole du Congo Beige 40: 109-20.

Charter, C. F. 1950 *Report on environmental conditions in the Potaro District, British Guiana, in relation to agricultural development, with special reference to cocoa production*. Kumasi, Ghana: Department of Soil and Land Use Survey (mimeo).

Charter, C. F. 1950/58 *Report on the environmental conditions prevailing in Block 'A', Southern Province, Tanganyika Territory*. Accra, Ghana: Department of Agriculture. (Original report mimeo, 1950; published, 1958).

Charter, C. F. 1957 The aims and objects of tropical soil surveys. *Soils and Fertilizers* 20:127-28.

Charter, C. F. 1957 *Suggestions for a classification of tropical soils*. Miscellaneous Paper 4. Kumasi, Ghana: Division of Soil and Land Use Survey.

Green, H. 1956 Mr C. F. Charter, O.B.E. [Obituary]. *Nature* 178: 346-47.

P. S. C. 1956 Mr C. F. Charter [Obituary]. *The Times* 6 February 1956, p. 12.

Charles Wright

For publications, see Chapters 9 and 11.

Hugh Brammer

Brammer, H. 1962 Soils. In *Agriculture and land use in Ghana* (ed. J. B. Wills, Oxford University Press), 88-126.

Brammer, H. 1967 *Soils of the Accra Plains. With map 1:125 000*. Memoir 3, Ghana Academy of Sciences. Kumasi, Ghana: Soil Research Institute.

Brammer, H. 1970 Some unsolved problems of soil science in East Pakistan. *Pakistan Journal of Soil Science* 6: 31-44.

Brammer, H. 1976 *Soils of Zambia*. Lusaka, Ministry of Rural Development. (With references to Brammer's other publications on Zambia.)

Brammer, H. 1996-2004 *The geography of the soils of Bangladesh* (1996); *Agricultural development possibilities in Bangladesh* (1997); *Agricultural disaster management in Bangladesh* (1999); *Agroecological aspects of ecological research in Bangladesh* (2000); *How to help small farmers in Bangladesh* (2002); *Land use and land use planning in Bangladesh* (2002); *Can Bangladesh be protected from floods?* (2004). Dhaka: University Press.

Anthony Smyth

[Brammer, H. and Coulter, J.] Tony Smyth (1927-2008) [Obituary]. *Agriculture for Development* 4: 32.

Maurice Purnell

Young, A. and others 2015 Maurice Purnell, 1930-2015 [Obituary]. *Agriculture for Development* 26: 54.

Distinguished guests: Stanislav Radwanski

Radwanski, S. A. 1960 *The soils and land use of Buganda: a reconnaissance survey*. Memoirs of the Research Division, Department of Agriculture. Series 1, Soils, No. 4. Entebbe: Government Printer.

Radwanski, S. A. 1969 Improvement of red acid sands by the neem tree (Azadirachta indica) in Sokoto, North-Western State of Nigeria. *Journal of Applied Ecology* 6: 507-11.

Radwanski, S. A. and Ollier, C. D. 1959 A study of an East African catena. *Journal of Soil Science* 10: 149-68.

Radwanski, S. A. and Wickens, G. E. 1967 The ecology of Acacia albida on mande soils in Zalingei, Jebel Marta, Sudan. *Journal of Applied Ecology* 4: 569-79.

Jan Kowal

Kowal, J. M. and Kassam, A. H. 1978 *Agricultural ecology of savanna*. Oxford, UK: Clarendon Press.

Chapter 7 Central and Southern Africa

Zambia (Northern Rhodesia)

Allan, W. 1949 *Studies in African land usage in Northern Rhodesia*. Rhodes-Livingstone Institute Paper 15. London: Rhodes-Livingstone Institute.

Allan, W. 1965 *The African husbandman*. Edinburgh, UK: Oliver and Boyd. New edition with an introduction by H. Tilley 2005, London: International African Institute.

Allan, W., Gluckman, M., Peters, D. U. and Trapnell, C. G. 1948 *Landholding and land usage among the Plateau Tonga of Mazabuka District; a reconnaissance survey, 1945*. Rhodes-Livingstone Paper 14. London: Cumberlege.

Astle, W. L. 1989 *South Luangwa National Park: landscape and vegetation*. Lusaka: Zambian Government.

Astle, W. L. 2005 A history of wildlife conservation and management in the mid-Luangwa valley. In *How green was our empire? Environment, development and the Colonial Service* (ed. T. Barringer, Institute of Commonwealth Studies, London), 75-103.

Astle, W. L., Webster, R. and Lawrence, C. J. 1969 Land classification for management planning in the Luangwa Valley of Zambia. *Journal of Applied Ecology* 57: 143-69.

Ballantyne, A. O. 1956 *Report of a soil and land-use survey of the Copperbelt, Northern Rhodesia*. Lusaka: Department of Agriculture.

Brammer, H. 1976 *Soils of Zambia*. Lusaka, Ministry of Rural Development. 2nd edition by W. J. Veldkamp 1987.

Dalal-Clayton, B. 1975 *The sandveldt soils of Central Province*. Soil Survey Report 32. Chilanga, Zambia: Mount Makulu Research Station.

FAO 1982 *Potential population supporting capacities of lands in the developing world*. Rome: FAO.

Priestley, M. S. J. W. and Greening, P. 1955 *Ngoni land utilisation survey 1954-55*. Lusaka: Government Printer.

Ruthenburg, H. 1971 *Farming systems in the tropics*. Oxford, UK: Clarendon.

Verboom, W. C. and Brunt, M. A. 1970 *An ecological survey of Western Province, Zambia*. Land Resources Study 8. Tolworth, UK: Land Resources Division.

Zimbabwe (Rhodesia, Southern Rhodesia)

Ellis, B. S. 1958 Soil genesis and classification. *Soils and Fertilizers* 21: 145-47.

Elwell, H. A. 1977 *A soil loss estimation system for Southern Africa*. Research Bulletin 22, CONEX, Salisbury.

Hudson, N. 1971 *Soil conservation*. London: Batsford.

Nyamapfene, K. 1991 *The soils of Zimbabwe*. Harare: Nehanda.

Thomas, R. G. and Ellis, B. S. 1955 *Provisional soil map of Southern Rhodesia 1:1 000 000*. Salisbury: Director of Federal Surveys. Second edition 1979.

Thomas, R. G. and Vincent, V. 1959 Classification of soils for land-use purposes in the Rhodesias. In *Proceedings of the Third Inter-African Soils Conference, Dalabar*, 351-60.

Thompson, J. G., Webster, R. and Young, A. 1960 *Soils map of the Federation of Rhodesia and Nyasaland, 1:2 500 000*. Federal Atlas Map 11. Salisbury: Federal Department of Surveys.

Vincent, V., Thomas, R. G. and Anderson, R. 1960 *An agricultural survey of Southern Rhodesia. Part 1: Agro-ecological survey. Part 2: The agro-economic survey*. Salisbury: Federation of Rhodesia and Nyasaland.

Lesotho (Basutoland)

Bawden, M. G. et al. 1968 *The land resources of Lesotho*. Land Resource Study 3. Tolworth, UK: Land Resources Division.

Binnie & Partners 1972 *Lesotho: study on water resources development*. Volumes 1-6. London: Binnie & Partners.

Carroll, D. M. and Bascomb, C. L. 1967 *Notes on the soils of Lesotho*. Technical Bulletin 1. Tolworth, UK: Land Resources Division. With soils map, 1:250 000.

Swaziland

Engineering and Power Development Consultants 1970 *Usutu, Mbuluzi, Momati and Lomati river basins: general plan for development and utilization of water resources*. Sidcup, UK: Engineering and Power Development Consultants.

Hunting Technical Services Ltd. 1961 *Swaziland irrigation scheme*. London: Hunting Technical Services Ltd.

Tons, J. H. 1967 *Veld types of Swaziland*. Bulletin 18. Swaziland: Department of Agriculture.

Tons, J. H. and Kidner, E. M. 1967 *Veld and pasture management in Swaziland – a guide for extension workers and farmers*. Bulletin 17. Swaziland: Department of Agriculture.

Murdoch, G. 1968 *Soils and land capability in Swaziland*. Mbabane: Ministry of Agriculture.

Murdoch, G. and Andriesse, J. 1964 *A soil and irrigability survey of the Lower Usutu Basin (South) in the Swaziland Lowveld*. London: HMSO. With maps scale 1:50 000.

Murdoch, G., Webster, R., and Lawrence, C. J. 1971 *Atlas of land systems in Swaziland*. Christchurch, UK: MEXE.

WOSSAC, n.d. *WOSSAC material held in the Archive for Swaziland*. www.wossac.com/archive/overview_swaziland.cfm

Botswana (Bechuanaland)

Bawden, M. G. (ed.) 1965 *Some soils of Northern Bechuanaland, with a description of the vegetation zones*. Tolworth, UK: Land Resources Division.

Bawden, M. G. et al. 1963 *The land resources of Eastern Bechuanaland*. Tolworth, UK: Forestry and Land Use Section, Directorate of Overseas Surveys.

Botswana: Ministry of Local Government and Lands 1977 *Lefatshe la rona – our land: the report on the Botswana Government public consultation on its policy proposals on tribal grazing lands*. Gaborone: Government Printer.

Langdale-Brown, I. et al. 1963 *Land use prospects of Northern Bechuanaland*. Tolworth, UK: Land Resources Division.

Mitchell, A. J. B. 1972 *The Central and Southern State Lands, Botswana*. Land Resource Study 11. Tolworth, UK: Land Resources Division.

Mitchell, A. J. B. 1976 *The irrigation potential of soils along the main rivers of Eastern Botswana: a reconnaissance assessment*. Land Resource Study 7. Tolworth: Land Resources Division.

George Murdoch

Baillie, I. 2005 George Murdoch's contribution to soils survey and agriculture in Swaziland. *Tropical Agriculture Association Newsletter* 25(3), September 2005, 28.

Landon, J. R. (ed.) 1984 *Booker tropical soil manual*. London: Longman.

Watson, J. 2005 Dr George Murdoch MBE – obituary. *Tropical Agriculture Association Newsletter* 25(2), June 2005, 28-29.

Chapter 8 Malawi (Nyasaland)

The Lower Shire Valley project

Halcrow, Sir William & Partners 1954 *Report on the control and development of Lake Nyasaland and the Shire River*. London: Halcrow.

Hunting Technical Services 1957 *The soil survey and land classification report of the Elephant Marsh area, Lower Shire Valley, Nyasaland*. London: Hunting Technical Services.

Ecological studies

Jackson, G. 1954 Preliminary ecological survey of Nyasaland. In: *Proceedings of the 2nd Inter-African Soils Conference, Leopoldville*. Brussels: Commission for Technical Cooperation in Africa South of the Sahara.

Jackson, G. 1961 *Climatic regions of Nyasaland*. Salisbury: Federal Department of Surveys.

Jackson, G. and Wiehe, P. O. 1958 *An annotated check list of Nyasaland grasses, indigenous and cultivated*. Zomba: Government Printer.

Williamson, J. 1956 *Useful plants of Nyasaland*. Zomba, Malawi: Government Printer. Second edition 1972 as *Useful plants of Malawi*.

The agro-ecological survey

Stobbs, A. R. 1971 *Malawi: natural regions and areas – environmental conditions and agriculture*. Sheet 3: Southern Malawi. Tolworth, UK: Directorate of Overseas Surveys.

Young, A. 1960 *Preliminary soil map of Nyasaland*. Zomba: Government Printer.

Young, A. 1961. Soil survey: Chikwawa-Ngabu. In *Annual Report of the Department of Agriculture, Nyasaland, Part II, 1959/60*, 191-201, with two maps.

Young, A. 1965 *Malawi: natural regions and areas – environmental conditions and agriculture. Sheet 1: Northern Malawi. Sheet 2: Central Malawi. Maps, scale 1:500 000*. Tolworth, UK: Directorate of Overseas Surveys.

Young, A. 1973. Soil survey procedures in land development planning. *Geographical Journal* 139, 53-64.

Young, A. and Brown, P. 1962. *The physical environment of northern Nyasaland, with special reference to soils and agriculture*. Zomba: Government Printer.

Young, A. and Brown, P. 1964. *The physical environment of central Malawi, with special reference to soils and agriculture*. Zomba: Government Printer

Land use

Stobbs, A. R. 1985 *Land use survey of Malawi 1965-67*. Surbiton, UK: Land Resources Development Centre.

Subsequent land resource survey

Kingston, J. D., Mitchell, A. J. B., Ntokotha, E. M., and Billing, D. W. n.d. *Soil surveys 1971-72*. Zomba: Government Printer.

Lowole, M. W. 1965 *Soil map of Malawi. 1:1 000 000. Draft*. Later drafts followed.

Geology, forests, soil fertility, and population

Chapman, J. D. and White, F. 1970 *The evergreen forests of Malawi*. Oxford: Commonwealth Forestry Institute.

Cooper, W. G. G. 1950 *The geology and mineral resources of Nyasaland*. Zomba: Government Printer. Revised edition 1957.

Dixey, F. 1925 *The physiography of the Shire Valley, Nyasaland, and its relation to soils, water supply and transport routes*. Zomba: Government Printer.

Dixey, F. 1926 The Nyasaland section of the Great Rift Valley. *Geographical Journal* 79: 117-140.

Dixey, F. 1955 *Some aspects of the geomorphology of Central and Southern Africa*. Alex L. du Toit Memorial Lectures 4. Geological Society of South Africa, Annexure to Volume 58.

Dixey, F. 1957 The East African rift system. *Colonial Geology and Mineral Resources*, Bulletin Supplement 1.

Dixey, F. 1957 Colonial geological surveys 1947-56. *Colonial Geology and Mineral Resources*, Bulletin Supplement 2.

Land husbandry and conservation agriculture

Dixey, F., Clements, J. B. and Hornby, A. J. W. 1924 *The destruction of vegetation and its relation to climate, water supply, and soil fertility*. Nyasaland Department of Agriculture, Bulletin 1.

Shaxson, T. F., Hunter, N. A., Jackson, T. R., and Alder, J. R. 1977 *A land husbandry manual*. Lilongwe: Ministry of Agriculture and Natural Resources.

Shaxson, T. F., Hudson, N. W., Sanders, D. W., Roose, E., and Moldenhauer, W. C. 1989 *Land husbandry: a framework for soil and water conservation*. Ankeny, USA: Soil and Water Conservation Society.

Chapter 9 The West Indies and Central America

The West Indies

For pre-1948 surveys (the 'Grey Books') see Hardy et al. In Chapter 2.

Brown, C. B. et al. 1965-1966 *Land capability survey: Trinidad and Tobago. Vols 1-5.* Trinidad: Government of Trinidad and Tobago.

Grove, R. H. 1995 *Green imperialism: Colonial expansion, tropical island Edens and the origins of environmentalism*, 1600-1860. Cambridge, UK: University Press.

Hardy, F. and Ahmad, N. 1974 Soil science at I.C.T.A./U.W.I., 1922-1972. *Tropical Agriculture* 51: 468-76.

Henry, P. W. T. 1974 *The pine forests of the Bahamas*. Land Resource Study 16. Tol worth, UK: Land Resources Division.

Jones, T. A., and 23 others, 1958-1971. *Soil and land-use surveys. Nos. 1-26*. St Augustine, Trinidad: Imperial College of Tropical Agriculture, and University of the West Indies. (the 'Green Books')

Little, B. G. et al. 1977 *Land resources of the Bahamas: a summary*. Land Resource Study 27. Tolworth, UK: Land Resources Division.

Shephard, C. Y. 1932 *The cocoa industry of Trinidad: some economic aspects. Series III. An examination of the effects of soil type on age and yield*. Port of Spain: Government Printer.

Belize (British Honduras)

King, R. B. et al. 1992 *Land resources assessment of Northern Belize*. Tolworth, UK: Natural Resources Institute.

Wright, A. C. S., Romney, D. H. et al. 1959 *Land in British Honduras: report of the British Honduras land use survey team*. Colonial Office Research Publication 24. London: HMSO.

Guyana (British Guiana)

FAO 1965. *Report on the soil survey project of British Guiana*. Vols I-VII. Rome: FAO.

Chapter 10 South Asia

India

Das, S. N. 2000 Soil information and land development. In *Advances in land resource management for the 21st Century* (ed. Secretary General ICLRM, Angkor Publishers, New Delhi), 51-62.

Ghosh, A. B. et al. 1984 *Soil science in India*. New Delhi: Indian Agricultural Research Institute.

Raychaudhuri, S. P. 1968 Development of soil and land use survey and soil tests in India for increased agricultural production. *Science and Culture* 34: 164-69.

Raychaudhuri, S. P. et al. 1963 *Soils of India*. New Delhi: Indian Council of Agricultural Research.

Sehgal, J. L. et al. 1990 *Agro-ecological regions of India*. NBSS Publication 24. New Delhi: National Bureau of Soil Survey and Land Use Planning.

Shokal'skaya Z. Y. 1932 *The natural conditions of soil formation in India*. USSR(?).

Uppal, B. N. et al 1953 *Final report of the All India soil survey scheme*. Delhi: Indian Council of Agricultural Research.

Satya Prasad Raychaudhuri

Biswas, T. D. and Gawande, S. P. 1990 Dr Satya Prasad Raychaudhuri – a biographical sketch. *Journal of the Indian Society of Soil Science* 38: 366-372.

Pakistan (West Pakistan)

FAO 1971 *Soil resources in West Pakistan and their development possibilities*. Rome: FAO.

Fraser, I. S. et al. 1958 *Report on a reconnaissance survey of the landforms, soils and present land use of the Indus plains, West Pakistan*. Ottawa: Government of Canada.

Glover, H. 1946 *Erosion in the Punjab*. Lahore: Feroz.

Hunting Technical Services Ltd 1961 *Mangla watershed management study*. Hemel Hempstead, UK: Hunting Technical Services Ltd.

Lander, P. E., Narain, R. and Lai, M. M.1929 *Soils of the Punjab*. Calcutta: Department of Agriculture.

Malcolm Macdonald and Partners, and Hunting Technical Services Ltd 1964-66 *Lower Indus Report*. London: Macdonald.

Roberts, W. and Karter Singh, S. B. S. 1951 *A textbook of Punjab Agriculture*. Lahore: Civil and Military Gazette. (See Chapters 4 and 5, Soil, and Chapter 8, Irrigation.

Punjab village surveys

Dass, A. ('under supervision of Calvert, H.') 1931 *Punjab village surveys 3. An economic survey of Tehong, a village in the Jullundur District of the Punjab*. Lahore: Board of Economic Enquiry.

Bangladesh (East Pakistan)

Brammer, H. 1996-2004 See Bibliography for Chapter 6.

FAO 1971 *Soil survey project, Bangladesh*. Technical Report 3. Soil resources. Rome: FAO.

Mookegee, D. N. 1909 *Notes on the soils of Bengal*. Calcutta: Bengal Secretariat Press.

Soil Resources Development Institute, Bangladesh 1967 *Reconnaissance soil survey of Dhaka District*. Dhaka: SRDI.

Soil Resources Development Institute, Bangladesh 1965-77 *Thirty-one District reconnaissance soil survey reports*. Dhaka: Soil resources Development Institute. (For details see Brammer, H. 1996 *The geography of the soils of Bangladesh*. Dhaka, University Press.)

Sri Lanka (Ceylon)

Ceylon: Survey Department 1962 *Ceylon: showing approximate distribution of great soil groups. Map, scale 1:506 880*. Colombo: Survey Department.

FAO 1969 *Mahaweli Ganga irrigation and hydropower survey, Ceylon: final Report*. Rome: UNDP.

Hunting Survey Corporation 1962 *A report on a survey of the resources of the Mahaweli Ganga Basin, Ceylon*. Toronto, Canada: Hunting Survey Corporation.

Joachim, A. W. R. 1945 The soils of Ceylon. *Tropical Agriculturalist* 111: 161-72.

Mapa, R. B. 2006 Landmarks of history of soil science in Sri Lanka. In *Abstracts, World Congress of Soil Science, Philadelphia, 2006*.

Moorman, F. R. and Panabokke, C. R. 1961 *Soils of Ceylon*. Colombo: Government Press.

Panabokke, C. R. 1996 *Soils and agro-ecological environments of Sri Lanka*. Colombo, Sri Lanka: Natural Resources, Energy and Science Authority.

Myanmar (Burma)

Charlton, J. 1932 *Soil survey – Mandalay Canal area, Burma. Agricultural Surveys 15.* Rangoon: Department of Agriculture.

FAO-Unesco 1977. *Soil map of the world. Vol. 7, South Asia.* Paris: Unesco.

See also Bruce (1923), Castens (1927) and Charlton (1935) listed under Early Scientific Accounts in Chapter 2.

Chapter 11 South-east Asia and the Pacific

West Malaysia (Malaya)

Coulter, J. K. 1964 Soil surveys and their application in tropical agriculture. *Tropical Agriculture* 41: 185-96.

Law Wei Min and Selvadurai, K. 1968 The 1968 reconnaissance soil map of Malaya. In *Proceedings of the 3rd Malaysian Soils Conference, Kuching,* 1968 (ed. J. P. Andriesse), 229-237.

Leamy, M. L. and Panton, W. P. 1966 *Soil survey manual for Malayan conditions.* Kuala Lumpur: Ministry of Agriculture and Cooperatives.

Lee Peng Choong and Panton, W. P. 1971 *First Malaysia plan: land capability classification report, West Malaysia.* Kuala Lumpur: Economic Planning Unit.

Malayan Soil Survey Reports 1958-70 *Reconnaissance soil survey.* A series of reports and maps for the States of Malaya. Authors include B. Gopinathan, D. W. Ives, Law Wei Min, W. P. Panton, T. R. Paton, S. Paramanathan, E. Pushparajah, J. R. D. Wall, I. F. T. Wong. Kuala Lumpur: Ministry of Agriculture and Lands.

Panton, W. P. 1958 *Reconnaissance soil survey of Trengganu.* Department of Agriculture Bulletin 105. Kuala Lumpur: Department of Agriculture.

Panton, W. P. 1964 The 1962 soil map of Malaya. *Journal of Tropical Geography* 18: 12-20.

Wyatt-Smith, J. and Panton, W. P. 1995 *Manual of Malayan silviculture of inland forest.* Kuala Lumpur: Forest Research Institute.

Land settlement projects: Jengka and Johor

Hunting Technical Services and Tippets-Abbett-McCarthy-Stratton 1967 *Jengka Triangle: Volume 1, The outline master plan. With Volume 2, text, and Volume 3, maps.*

Hunting Technical Services, Binnie & Partners, Overseas Development Group of the University of East Anglia, and Shankland Cox Overseas 1971 *Johor Tenggara regional master plan. Volumes 1-9 plus map folder.*

East Malaysia: Sarawak (Sarawak)

J. P. Andriesse 1966 *Soil and land potential of Kuching-Bau-Lundu area.* Kuching: Department of Agriculture.

Anon [Keresa Plantation] 1978 *Soil suitability for oil palm cultivation of an area of 2916.8 acres of Sarawak Oil Palm Sendirian Berhad.* Kajang, Selangor, Malaysia: Agricultural Research and Advisory Bureau.

Paramananthan, S. 1974 *Semi-detailed soil survey of the Sungei Tunoh and Sungei Pila areas, Seventh Division, Sarawak.* Kuala Lumpur, Malaysia: Government of Malaysia.

Wall, J. R. D. 1962 *Report on the reconnaissance soil survey of Bekenu-Niah-Suai area.* Kuching: Department of Agriculture. (One of a series of reconnaissance soil surveys 1962-66.

East Malaysia: Sabah (North Borneo)

Acres, B. D. et al. 1975 *The soils of Sabah.* Volumes 1-5 with map folder. Land Resource Study 20. Tolworth, UK: Land Resources Division.

Clare, K. E. and Beaven, P. J. 1965 *Roadmaking materials in Northern Borneo.* Road Research Technical Paper 68. London: HMSO.

Paton, T. R. 1963 *A reconnaissance soil survey of the Semporna Peninsula, North Borneo.* Colonial Research Studies 36. London: HMSO.

Thomas, P., Lo, F. C. K. and Hepburn, A. J. 1976 *The land capability classification of Sabah. Volumes 1-4 with map folder.* Land Resource Study 25. Tolworth, UK: Land Resources Division.

Brunei

Blackburn, G. et al. 1958 *A soil survey of part of Brunei, British Borneo.* Melbourne, Australia: CSIRO.

Hunting Technical Services 1969 *Land capability study [Brunei]. Volumes 1-3.* Borehamwood, UK: Hunting Technical Services Ltd.

Samoa (Western Samoa)

Wright, A. C. S. 1963 *Soils and land use of Western Samoa.* Soil Bureau Bulletin 22. Wellington: New Zealand Soil Bureau.

Fiji

Berry, M. J. et al. 1973 *Fiji forest inventory.* Land Resource Study 12. Tolworth: Land Resources Division.

Twyford, I. T. and Wright, A. C. S. 1965 *The soil resources of the Fiji Islands. Volume I text, Volume II map box.* Suva, Fiji: Government Printer.

Wright, C. H. 1916 Fiji: report on the soils. *Fiji Agricultural Bulletin* 9.

Papua New Guinea

Bleeker, P. et al. 1975 *Explanatory notes to the land limitation and agricultural land use potential map of Papua New Guinea. With map at 1:1000 000.* Land Research Series 36. Melbourne, Australia: CSIRO.

Bleeker, P. et al. 1989 *Explanatory notes to the soils map of Papua New Guinea. With map at 1:1 000 000.* Natural Resources Series 10. Canberra, Australia: Division of Water and Land Resources.

Chartres, C. J. 1981 Land resources assessment for sugar cane cultivation in Papua New Guinea, *Journal of Applied Geography* 1: 259-273.

Dearden, P. N., Freyne, D. F. and Humphreys, G. S. 1986 Soil and land resource surveys in Papua New Guinea. *Soil Survey and Land Evaluation* 6: 43-50.

Haantjens, H. A. et al. 1964 *Lands of the Buna-Kokoda area, Territory of Papua New Guinea.* Land Research Series 20. Melbourne, Australia: CSIRO. (Similar surveys of areas of Papua New Guinea form Land Research Series 14, 17, 27, 29, 31, 1965-70.)

Holloway, R. S. et al. 1973 *Land resources and agricultural potential of the Markham Valley.* Research Bulletin 14 (Parts 1-10). Port Moresby: Department of Primary Industry.

Humphreys, G. S. 1998 A review of some important soil studies in Papua New Guinea. *Papua New Guinea Journal of Agriculture, Forestry and Fisheries* 41(1): 1-19.

Löffler, E. 1974 *Explanatory notes to the geomorphological map of Papua New Guinea. With map at 1:1 000 000.* Melbourne, Australia: CSIRO.

Paijmans, K. 1975 *Explanatory notes to the vegetation map of Papua New Guinea.* Melbourne, Australia: CSIRO.

Pacific islands

Grange, L. I. et al. 1953 *Soils of the Lower Cook Group.* Soil Bureau Bulletin 8. Wellington, New Zealand: Soil Bureau.

Hansell, J. R. F. et al. 1974 *Land resources of the Solomon Islands*. Land Resource Study 18. 8 volumes and 20 supplementary reports. Tolworth, UK: Land Resources Division.

Johnson, M. S. 1971 *New Hebrides Condominium, Erromango forest inventory*. Land Resource Study 10. Tolworth, UK: Land Resources Division.

Orbell, G. E. 1971 *Soil survey: Vav'u and adjacent islands, Tonga Islands*. Bulletin 8. Wellington, New Zealand: Royal Society of New Zealand.

Quantin, P. 1982 *Vanuatu: agronomic potential and land use map*. Paris: ORSTOM.

Twyford, I. T. 1958 *Pitcairn island*. With 6 maps at 1:11 520. Mimeo.

UNCED 1992 *Agenda 21: a programme of action for sustainable development*. Rio de Janeiro, Brazil: UN Conference on Environment and Development.

Other island territories

Brenna, S. 1946 *Soil map of the Falkland Elands*. London: Admiralty.

Grant, C. J. 1960 *The soils and agriculture of Hong Kong*. Hong Kong: Government Press.

Jenkin, R. N. et al. 1968 *An investigation into the coconut-growing potential of Christmas Island*. Land Resource Study 4. Tolworth, UK: Land Resources Division.

Lang, D. M. 1960 *Soils of Malta and Gozo*. Colonial Research Studies 29. London: HMSO.

Piggott, C. J. 1968 *A soil survey of Seychelles*. Technical Bulletin 2. Tolworth, UK: Land Resources Division.

Chapter 12 Maps, rocks, climate, plants, and land use

General

Bunting, A. H. (ed.) 1987 *Agricultural environments: characterization, classification and mapping*. Wallingford, UK: CAB International.

Maps: topographic survey

Macdonald, A. 1996 *Mapping the world: a history of the Directorate of Overseas Surveys 1946-1985*. London: HMSO.

McGrath, G. 1983 Mapping for development: the contributions of the Directorate of Overseas Surveys. Monograph 29-30. *Cartographica* 20: 1-264. With 10 map extracts.

Rocks: geological survey

Dixey, F. 1926 The Nyasaland section of the Great Rift Valley. *Geographical Journal* 79: 117-140.

Dixey, F. 1926 *The East African rift system*. Colonial Geology and Mineral Resources, Bulletin Supplement 1.

Dixey, F. 1957 *Colonial geological surveys 1947-56*. Colonial Geology and Mineral Resources, Bulletin Supplement 2. London: HMSO.

Climate and soil water resources

Braun, H. M. H. The agro-climatic zone map of Kenya. In *Exploratory soil map and agro-climatic zone map of Kenya, 1980, scale 1:1 000 000* (W. G. Sombroek, H. M. H. Braun and B. J. A. van der Pouw, Ministry of Agriculture, Nairobi), 43-52.

FAO 1978 *Report on the agro-ecological zones project. Vol. 1: Methodology and results for Africa*. World Soil Resources Report 48. Rome: FAO.

Kowal, J. M. and Kassam, A. H. 1978 *Agricultural ecology of savanna*. Oxford, UK: Clarendon Press.

Penman, H. L. 1948 Natural evaporation from open water, bare soil and grass. *Proceedings of the Royal Society, Series A* 193: 120-45.

Pereira, C. 1973 *Land use and water resources*. Cambridge, UK: University Press.

Pereira, C. 2000 *Simama: a lifetime study of tropical issues*. Maidstone, UK: Teston Books.

Pereira, Sir Charles. Obituary. *The Times* 13 January 2005.

Pereira, Sir Charles. Obituary. *Tropical Agriculture Association Newsletter* 25(1), March 2005, 30-31.

Pratt, D. J., Greenway, P. J. and Gwynne, M. D. 1966 A classification of East African rangeland with an appendix on terminology. *Journal of Applied Ecology* 3: 369-382. (Includes agro-climatic zones)

Plants: ecological survey and vegetation mapping

Blair-Rains, A. et al. 1979 Mapping Nigeria's vegetation from radar: discussion. *Geographical Journal* 145: 274-81

Brunt, M. and Trapnell, C. J. 1986 *Vegetation and climate maps of South-Western Kenya. 8 map sheets*. Tolworth, UK: Land Resources Development Centre.

Kuchler, A. W. 1956 Classification and purpose in vegetation maps. *Geographical Review* 46: 155-67.

Moss, R. P. 1968 Land use, vegetation and soil factors in south west Nigeria, a new approach. *Pacific Viewpoint* 9: 107-26.

Trapnell, C. J. See Bibliography for Chapter 2.

Pasture resource survey

FAO 1991 *Guidelines: land evaluation for extensive grazing.* FAO Soils Bulletin 58. FAO: Rome.

Pratt, D. J. and Gwynne, M. D. 1977 *Rangeland management and ecology in East Africa.* London: Hodder and Stoughton.

Forest inventory

FAO 1984 *Land evaluation for forestry.* Forestry Paper 48. Rome: FAO.

Young, A. 1993. Land evaluation and forestry management. In *Tropical forestry handbook* (ed. L. Pancel, Springer, Berlin), 811-45. Revised version in *Tropical Forestry Handbook, Second Edition* (ed. L. Pancel and M. Köhl, Springer, Berlin), 1835-1868.

Land use

Alexandratos, N. (ed.) 1995 *World agriculture: towards 2010.* Chichester, UK: Wiley, for FAO.

Bruinsma, J. (ed.) 2003 *World agriculture: towards 2015/2030.* London: Earthscan, for FAO.

Hornby, A. J. W. 1935 *Central Nyasaland: preliminary map of land utilisation. Scale 1:500 000.* Zomba: Government Printer.

Young, A. 1994. Towards international classification systems for land use and land cover. In *Report of the UNEP/FAO expert meeting on harmonizing land cover and land use classifications, Geneva, 23-25 November 1993*, GEMS Report Series 25, UNEP and FAO, Nairobi, Annex V, 44 pp.

Young, A. 1995 Competition for land. Chapter 4 in *Land resources: now and for the future.* Cambridge, UK: University Press.

Is there really spare land?

Young, A. 1999. Is there really spare land? A critique of estimates of available cultivable land in developing countries. *Environment, Development and Sustainability* 1: 3-18.

Young, A. and Nachtergaele, F. O. 2003. Potential and constraints of soils for increased agricultural production: how much spare land? In *People matter: food security and soils* (ed. R. Lahmar, M. Held and L. Montanarella, Torba Soil and Society, Montpellier, France), 21-31.

Chapter 13 From soil conservation to conservation agriculture

Awareness of erosion

Ainslie, S. R. 1935 *Soil erosion in Kenya*. Kaduna, Nigeria: Government Printer.

Cox, P. et al. 1935 The encroaching Sahara: the threat to the West African colonies: discussion. *Geographical Journal* 85: 519-24.

Dixey, F., Clements, J. B. and Hornby, A. J. W. 1924 *The destruction of vegetation and its relation to climate, water supply, and soil fertility*. Department of Agriculture Bulletin 1. Zomba: Government Printer.

Fuchs, V. E. 1931 *Report of Cambridge scientific expedition to the East African lakes, 1930-31*. Quoted by Huxley (1938), q.v. below, p.1061.

Glover, H. M. 1946 *Erosion in the Punjab, its causes and cure: a survey of soil conservation*. Lahore, India [now Pakistan]: Feroz.

Hailey, Lord 1938 *An African survey*. London: Oxford University Press.

Hardy, F. 1942 Soil erosion in Trinidad and Tobago. *Tropical Agriculture* 29: 29-35.

Hornby 1934 *Denudation and soil erosion in Nyasaland*. Department of Agriculture Bulletin 11 (New Series). Zomba: Government Printer.

Huxley, E. 1938 *Soil erosion [in Africa]*. In Hailey (1938), q.v., pp. 1056-1113. (Many early sources, not listed here, are given in footnotes.)

Jacks, G. V. and Whyte, R. O. 1938 *Erosion and soil conservation*. Technical Communication 36. Rothamsted, UK: Imperial Bureau of Soil Science.

Pole-Evans, I. B. 1939 *Report on a visit to Kenya*. Nairobi: Government Printer.

Stockdale, F. 1937 Soil erosion in the Colonial Empire. *Empire Journal of Experimental Agriculture* 5: 290-308.

Conservation by earth structures

Hudson, N. 1971 *Soil conservation*. London: Batsford.

Sheng, T. C. 1986 *Watershed conservation: a collection of papers for developing countries*. Taipei, Taiwan: Chinese Soil and Water Conservation Society.

Showers, K. B. 2005 *Imperial gullies: soil erosion and conservation in Lesotho*. Ohio, USA: Ohio University Press.

Wenner, C. G. 1980 *An outline of soil conservation in Kenya*. Nairobi: Ministry of Agriculture.

Forerunners of a new approach

Tempany, H. A. 1949 *The practice of soil conservation in the British Colonial Empire*. Technical Communication 45. Harpenden, UK: Commonwealth Bureau of Soil Science.

Conservation agriculture

Roose, E. 1996 *Land husbandry: components and strategy*. FAO Soils Bulletin 70.

Shaxson, T. F. (ed.) 1977 *A land husbandry manual*. Lilongwe, Malawi: Ministry of Agriculture and Resources.

Shaxson, T. F. 1980 Developing concepts of land husbandry for the tropics. In *Soil conservation: problems and prospects* (ed. R. P. C. Morgan, Wiley), 351-62.

Shaxson, T. F., Hudson, N. W., Sanders, D. W., Roose, E., and Moldenhauer, W. C. 1989 *Land husbandry: a framework for soil and water conservation*. Ankeny, USA: Soil and Water Conservation Society.

WOCAT: World overview of conservation agriculture approaches & technologies. www.wocat.net

Young, A. 1974 *Slopes* (pages 68-74).

Agroforestry and conservation agriculture

Kiepe, P. 1995 *No runoff, no soil loss: soil and water conservation in hedgerow barrier systems*. Tropical Resource Management Papers 10. Wageningen, Netherlands: Agricultural University.

Leaky, D. G. B. 1949 Changes in systems of cultivation aimed at limiting soil degradation by development of the cultivation of perennial tree crops in the Central Province of Kenya Colony. *Bulletin Agricole du Congo Belge* 40: 2164-72.

Young, A. 1989 *Agroforestry for soil conservation*. Wallingford, UK: CAB International. (Translated into French and Chinese.)

Young, A. 1997 *Agroforestry for soil management*. Wallingford, UK: CAB International.

Land degradation

Bai, Z. G., Dent, D. L., Olsson, and Schaepman, M. E. 2008 Proxy global assessment of land degradation. *Soil Use and Management* 24: 223-34. (With references to major outputs of the GLADA project.)

FAO *Land degradation assessment in drylands (LADA)*. www.fao.org/nr/lada

Oldeman, L. R., Hakkeling, R. T. A., and Sombroek, W. G. 1990 *World map of the status of human-induced soil degradation*. Wageningen, Netherlands: World Soil Information (ISRIC).

Young, A. 1994 *Land degradation in South Asia: its severity, causes, and effect upon the people*. World Soil Resources Report 78, UNDP/UNEP/FAO, Rome.

Young, A. 1995 Land degradation. Chapter 7 in *Land resources: now and for the future*. Cambridge, UK: University Press.

Chapter 14 Retrospect: the surveyors

This chapter is based on Colonial Office staff lists, information on the Land Resources Division coming from several sources, and above all, the details of education and careers sent by many correspondents, as recorded in the text.

Chapter 15 Retrospect: the surveys

Applications: actual and potential uses of land resource surveys

Brammer, H. 2002 *Land use and land use planning in Bangladesh*. Dhaka: University Press.

Charter, C. F. 1957 The aims and objectives of tropical soil surveys. *Soils and Fertilizers* 20: 127-28.

Chenery, E. M. 1960 *An introduction to the soils of the Uganda Protectorate*. Department of Agriculture, Uganda, Memoirs of the Research Division, Series I, Soils, No. 1. Kampala, Uganda Government.

Coulter, J. K. 1964 Soil surveys and their application in tropical agriculture. *Tropical Agriculture* 41: 185-96.

Dalal-Clayton, B. and Dent, D. 1993 *Surveys, plans and people: a review of land resource information and its use in developing countries*. Environmental Planning Issues 2. London: International Institute for Environment and Development.

Dalal-Clayton, B. and Dent, D. 2001 *Knowledge of the land: land resources information and its use in rural development*. Oxford, UK: University Press.

Young, A. 1973 Soil survey procedures in land development planning. *Geographical Journal* 139: 53-64.

Young, A. 1998 *Land resources: now and for the future.* Cambridge, UK: University Press.

Soil survey in developing countries today: soil institutions

FAO

www.fao.org/soils-portal/soil-survey/soil-maps-and-databases/national-soil-information-centers

Awareness and attitudes

International Union of Soil Sciences.
www.iuss.org

Chapter 16 The study of soils in the field: what is its role today?

World Soil Survey and Catalogue Archive.
www.wossac.com.archive/holdings.cfm

Soil health monitoring

Young, A. 1991 Soil monitoring: a basic task for soil survey organizations. *Soil Use and Management* 7: 126-30.

Shepherd, K. D., Shepherd, G. and Walsh, M. G. 2015 Land health surveillance and response: a framework for evidence-informed land management. *Agricultural Systems 132: 93-106.*

The conservation agriculture adviser

See sources for Conservation Agriculture in Chapter 13.

The integral nature of population policy

Kettlewell, R. W. 1955 *Outline of agrarian problems and policy in Nyasaland.* Legislative Council Paper, Government Printer, Zomba, Nyasaland [Malawi].

United Nations 1994 *Program of action adopted at the International Conference on Population and Development.* New York, United Nations Fund for Population Activities, 1994. *Twentieth Anniversary Edition 2004* www.unfpa.org/publications/international-conference-population-and-development-programme-action

Young, A. 2005 Poverty, hunger and population policy: linking Cairo with Johannesburg. *Geographical Journal* 171: 83-95.

LIST OF ACRONYMS

Titles of organizations are as they were at the time referred to in the text.

ACIU, Allied Central Interpretation Unit, Medmenham, UK

AISLUS, All India Soil and Land Use Survey

ASSOD, Status of Human-Induced Soil Degradation in South and Southeast Asia

BRALUP, Bureau of Resource Assessment and Land Use Planning (Tanzania)

CABI, Commonwealth Agricultural Bureau International

CCTA, Commission for Technical Cooperation in Africa

CDC, Commonwealth Development Corporation

CGIAR, Consortium of International Agricultural Research Centers

CONEX, Department of Conservation and Extension (Zimbabwe)

CSIR, CSIR Soil Research Institute (Ghana)

CSIRO, Commonwealth Scientific and Industrial Research Organisation (Australia)

DFID, Department for International Development (UK)

DOS, Directorate of Overseas Surveys (UK)

DTA, Diploma in Tropical Agriculture (Trinidad)

EAAFRO, East African Agriculture and Forestry Research Organisation

FELDA, Federal Land Development Authority (Malaysia)

GLADA, Global Assessment of Land Degradation and Improvement

GLASOD, Global Assessment of Human-induced Soil Degradation

IBSRAM, International Board for Soil Research and Management

ICAR, Indian Council of Agricultural Research

ICRAF, International Centre for Research in Agroforestry

ICRISAT, International Crops Research Institute for the Semi-Arid Tropics

ICTA, Imperial College of Tropical Agriculture (Trinidad)

IFPRI, International Food Policy Research Institute

IIED, International Institute for Environment and Development

IITA, International Institute of Tropical Agriculture

ILCA, International Livestock Centre for Africa

INRA, Institut national de la recherche agronomique

IRRI, International Rice Research Institute

ISRIC, International Soil Reference and Information Centre (Netherlands)

ITC, International Training Centre for Aerial Survey

IUFRO, International Union of Forest Research Organizations

IWMI, International Water Management Institute

KENSOTER, Kenya Soil and Terrain Database

LADA, Land Degradation Assessment in Drylands

LIFDC, Low income food-deficit country

LRD, Land Resources Division (UK)

MEXE, Military Engineering Experimental Establishment

NCAP, National Collection of Aerial Photography (UK)

NDVI, Normalized Difference Vegetation Index

NIRAD, Nigerian Radar Project

NOAA, National Oceanic and Atmospheric Administration (USA)

ORSTOM, Office de la Recherche Scientifique et Technique Outre-mer (France)

RRC, Regional Research Centre (West Indies)

SCARP, Special Conservation and Reclamation Project (Pakistan)

SLAR, Sideways Looking Airborne Radar

SLEMSA, Soil Loss Estimation Model for Southern Africa

SRICANSOL, Sri Lankan-Canadian Soil Resource Information System for Land Evaluation

STIBOKA, Stichting voor Bodemkartering (Foundation for Soil Mapping (Netherlands)

TAMS, Tippetts-Abbett-Mc-Carthy-Stratton (USA)

TANRIS, Tanzania Natural Resources Information System

UDI, Unilateral Declaration of Independence (Rhodesia)

ULG, ULG Consultants Ltd. (UK)

UNCED, United Nations Conference on Environment and Development

UNDP, United Nations Development Programme

UNEP, United Nations Environment Programme

UNFPA, United Nations Fund for Population Activities

USAID, United States Agency for International Development

USLE, Universal Soil Loss Equation

UWI, University of the West Indies

VSO, Voluntary Service Overseas (UK)

WOSSAC, World Soil Survey Archive and Catalogue (Cranfield University, UK)

INDEX OF PERSONAL NAMES

GENERAL INDEX

Made in the USA
Columbia, SC
23 November 2017